AF595031

End-Users' Perspectives on Energy Policy and Technology

End-Users' Perspectives on Energy Policy and Technology

Editor

Sung-Yoon Huh

MDPI • Basel • Beijing • Wuhan • Barcelona • Belgrade • Manchester • Tokyo • Cluj • Tianjin

Editor
Sung-Yoon Huh
Seoul National University of
Science & Technology
Korea

Editorial Office
MDPI
St. Alban-Anlage 66
4052 Basel, Switzerland

This is a reprint of articles from the Special Issue published online in the open access journal *Energies* (ISSN 1996-1073) (available at: https://www.mdpi.com/journal/energies/special_issues/End-Users_Perspectives_on_Energy_Policy_and_Technology).

For citation purposes, cite each article independently as indicated on the article page online and as indicated below:

LastName, A.A.; LastName, B.B.; LastName, C.C. Article Title. *Journal Name* **Year**, *Volume Number*, Page Range.

ISBN 978-3-0365-0015-7 (Hbk)
ISBN 978-3-0365-0016-4 (PDF)

Contents

About the Editor

Sung-Yoon Huh is an Assistant Professor in the Department of Energy Policy at Seoul National University of Science & Technology, where he has been a faculty member since 2017. Dr. Huh completed his Ph.D. at Seoul National University and his undergraduate studies at the same institute. He also went through two years of postdoctoral studies at the University of California at Berkeley. His research interests lie in the area of energy and environmental economics, ranging from theory to actual policy practice. Dr. Huh has published on a wide range of topics, including economic valuation, social acceptance, and innovation diffusion of energy (also environmental) projects and technologies. His articles have appeared in renowned international journals such as *Renewable and Sustainable Energy Reviews*, *Applied Energy*, *Energy Policy*, *Energy Economics*, *Sustainability*, and *Energies*. He has also actively collaborated with researchers in several other disciplines to provide better insights into the energy sector.

Preface to "End-Users' Perspectives on Energy Policy and Technology"

The global energy market is changing rapidly, and several megatrends can be identified in its transition. Among them, with a wider application of new energy technologies such as renewable energy, there is a possibility of a shift from a conventional centralized energy supply system to a more distributed energy production system. In addition, as people's overall economic level, education level, and living standards have improved, public interest and participation in energy policy and technology are steadily increasing. In many countries, the rejection of several energy projects due to the opposition of local residents is a representative example of such importance of public opinion in the process of energy policy implementation. In such a situation, it is very important to understand the social needs and public preferences for energy policy and technology and reflect them fully in future energy policy and technology development. In this Special Issue, differents kinds of theoretical and empirical studies analyzing the general public's perceptions and behavior regarding new energy technology and policy are presented. Through these papers, it is possible to predict future changes in the energy market as well as in the behavior patterns of end-users.

Sung-Yoon Huh
Editor

Article

A Machine Learning Solution for Data Center Thermal Characteristics Analysis

Anastasiia Grishina [1], Marta Chinnici [2,*], Ah-Lian Kor [3], Eric Rondeau [4] and Jean-Philippe Georges [4]

[1] Department of Mathematics and Computer Science, Eindhoven University of Technology, 5612 AZ Eindhoven, The Netherlands; a.grishina@tue.nl
[2] Department of Energy Technologies and Renewable Sources, ICT Division, ENEA Casaccia Research Center, 00123 Rome, Italy
[3] School of Built Environment, Engineering and Computing, Leeds Beckett University, Leeds LS1 3HE, UK; a.kor@leedsbeckett.ac.uk
[4] Université de Lorraine, Centre National de la Recherche Scientifique (CNRS), Centre de Recherche en Automatique de Nancy (CRAN), F-54000 Nancy, France; eric.rondeau@univ-lorraine.fr (E.R.); jean-philippe.georges@univ-lorraine.fr (J.-P.G.)
* Correspondence: marta.chinnici@enea.it; Tel.: +39-349-3553790

Received: 13 July 2020; Accepted: 20 August 2020; Published: 25 August 2020

Abstract: The energy efficiency of Data Center (DC) operations heavily relies on a DC ambient temperature as well as its IT and cooling systems performance. A reliable and efficient cooling system is necessary to produce a persistent flow of cold air to cool servers that are subjected to constantly increasing computational load due to the advent of smart cloud-based applications. Consequently, the increased demand for computing power will inadvertently increase server waste heat creation in data centers. To improve a DC thermal profile which could undeniably influence energy efficiency and reliability of IT equipment, it is imperative to explore the thermal characteristics analysis of an IT room. This work encompasses the employment of an unsupervised machine learning technique for uncovering weaknesses of a DC cooling system based on real DC monitoring thermal data. The findings of the analysis result in the identification of areas for thermal management and cooling improvement that further feeds into DC recommendations. With the aim to identify overheated zones in a DC IT room and corresponding servers, we applied analyzed thermal characteristics of the IT room. Experimental dataset includes measurements of ambient air temperature in the hot aisle of the IT room in ENEA Portici research center hosting the CRESCO6 computing cluster. We use machine learning clustering techniques to identify overheated locations and categorize computing nodes based on surrounding air temperature ranges abstracted from the data. This work employs the principles and approaches replicable for the analysis of thermal characteristics of any DC, thereby fostering transferability. This paper demonstrates how best practices and guidelines could be applied for thermal analysis and profiling of a commercial DC based on real thermal monitoring data.

Keywords: data center; thermal characteristics analysis; machine learning; energy efficiency; clustering; unsupervised learning

---

1. Introduction

Considerable efforts have been made by Data Centers in terms of their energy efficiency, reliability and sustainable operation over the past decade. A continuous increase in computing and power demands has spurred DCs to respond and upgrade their facilities in terms of size and stability [1,2]. A rapid growth of the information technology (IT) industry, advent of IoT, and AI technologies requires an exponentially expanding amount of data to be stored and processed. Consequently, smart DC management is on the rise to meet this demand. If a data center experiences a system failure or outage, it becomes challenging to ensure a stable and continuous provision of IT services, particularly for smart businesses, social media, etc. If such a situation occurs on a large scale, it could be detrimental to the businesses and public sectors that rely on DC services, for example, health systems, manufacturing, entertainment, etc. In other words, a data center has emerged as a mission-critical infrastructure [3] to the survival of public and business sectors enabled by smart technologies. Therefore, it warrants an exceptional necessity for the backup system management and uninterruptible power supply (UPS) systems so that computing system stability can be maintained even in emergency situations. Overall, thermal management involves the reduction of excess energy consumption by cooling systems, servers, and their internal fans. This encompasses the compliance of the IT room environment to the requirements stipulated in IT equipment specifications and standards that ensure better reliability, availability, and overall improved IT equipment performance.

The mission-critical facility management for the stable operation of a DC leads to huge cost increases, and careful reviews must be performed starting from the initial planning stage [4,5]. Moreover, IT servers require uninterruptible supplies of not only power but also cooling [6,7]. Undeniably, a significant increase in power density has led to a greater cooling challenge [8]. For this purpose, the design of a central cooling system in a liquid cooling architecture includes cooling buffer tanks. This design provides chilled water supply in emergency situations. During cooling system outages, the interruption of chillers triggers activation of emergency power and cooling supplies to restore cooling services. However, such emergency situations are infrequent on the scale of a DC life-cycle. In addition, recent the development of IT equipment has increased servers' tolerance to various operational challenging conditions compared to that in the past. Consequently, the operating times and capacities of chilled water storage tanks could be considerably reduced. The same is true for energy and water consumption of the liquid cooling system. Undeniably, it is imperative to explore the thermal specifications of the DC IT equipment. They are expressed as (but not limited to) different admissible ranges for temperature, humidity, periods of overheating before automatic power off, etc.

Given that IT devices might have different recommended specifications for their operation, maintaining healthy operational conditions is a complex task. Undoubtedly, covert thermal factors tend to affect the health of IT and power equipment systems in a negative way. Such factors comprise recirculation, bypass, and (partial) server rack overheating and stable operation is critical for a DC [9]. For example, in the case where an IT room is divided into cold and hot aisles, ineffective partitioning of the aisles (e.g., poor physical separation of the aisles) may result in leaked air causing recirculation of hot air (note: this is the mixing of hot and cold air) or cold air bypass (note: this happens if the cold air passes by the server and does not cool it properly - this occurs when the speed of air-flow is too high or the cold air is unevenly directed at the hot servers) [10]. Consequently, such emerging challenges have to be addressed to effect thermal conditions optimization within a DC facility. Undeniably, an increase in ambient temperature could lead to an increase in power usage of IT equipment that could lead to hardware degradation [8]. Thus, it is necessary to address the issue of server waste heat dissipation with the ultimate goal to attain an even thermal distribution within a premise. This is possible by taking appropriate measures to prevent the emergence of heat islands that result in individual server overheating [11].

This work explores IT room thermal characteristics using data mining techniques for the purpose of relevant and essential knowledge discovery. The primary goal is to use an unsupervised machine learning technique to uncover inadequacies in the DC cooling system based on real monitored thermal data. Analysis in this research leads to the identification of areas for improved thermal management and cooling that will feed into DC recommendations. The proposed methodology includes statistical analysis of IT room thermal characteristics and identification of individual servers that are contributors to hotspots (i.e., overheated areas) [12]. These areas emerge when individual servers do not receive adequate cooling. The reliability of the analysis has been enhanced due to the availability of a dataset of ambient air temperature in the hot aisle of an ENEA Portici CRESCO6 computing cluster [9].

In brief, clustering techniques have been used for hotspots localization as well as server categorization based on surrounding air temperature ranges. The principles and approaches employed in this work are replicable for thermal analysis of DC servers and thus, foster transferability. This work showcases the applicability of best practices and guidelines in the context of a real commercial DC that transcends the typical set of existing metrics for DC thermal characteristics assessment. Overall, this paper aims to raise DC thermal awareness and formulate recommendations for enhanced thermal management. This aim is supported by the following list of research objectives [9]:

RO.1.To identify a clustering (grouping) algorithm that is appropriate for the purpose of this research;
RO.2.To determine the criteria for feature selection in the analysis of DC IT room thermal characteristics;
RO.3.To determine the optimal number of clusters for the analysis of thermal characteristics;
RO.4.To perform sequential clustering and interpretation of results for repeated time series of air temperature measurements;
RO.5.To identify servers that most frequently occur in cold or hot air temperature ranges (and clusters);
RO.6.To provide recommendations related to IT room thermal management with the aim of appropriately addressing servers overheating issue.

In summary, typical identification of hotspots (and rack or cluster failure) in data centers is through the deployment of heatmaps (e.g., by Facebook in [13]) which comprise discrete scene snapshots of the premise (and not individual compute node) under study. However, our novel contribution is the employment of an Artificial Intelligence approach (i.e., Machine Learning clustering technique) based on 'continuous' data center environmental monitoring data, which will provide insights into the qualitative measure of the 'duration' a particular compute node is in a particular temperature range (i.e., low, medium, hot). Additionally, our proposed approach has an edge over typical heatmaps due to the low level granularity (i.e., with more details) information it provides for each node rather than aggregated information of the nodes (in clusters), so that targeted as well as effective corrective action or interventions can be appropriately taken by data center owners.

The remainder of the paper is organized as follows: Section 2 focuses on the background of the problem and related work; Section 3 presents the methodology used in this work; Section 4 provides discussion of experimental results and analysis; Section 5 concludes the paper with a summary and recommendation for future work.

2. Background and Related Work

In recent years, a number of theoretical and practical studies have been conducted on DC thermal management to better understand ways to mitigate inefficiencies of the cooling systems. This includes DC thermal and energy performance evaluation and load distribution optimization. Ineffective thermal management could be the primary contributor to DC IT infrastructure unreliability due to hardware degradation.

Existing DC-related thermal management research highlights the primary challenges of cooling systems in high power density DCs [14]; recommends a list of thermal management strategies based on energy consumption awareness [2,15]; explores the effect of different cooling approaches on power usage effectiveness (PUE) using direct air with a spray system that evaporates water to cool and humidify incoming air [16]; investigates the thermal performance of air-cooled data centers with raised and non-raised floor configurations [17]; studies various thermofluid mechanisms using cooling performance metrics [18]; proposes thermal models for joint cooling and workload management [19], while other strains of research explore thermal-aware job scheduling, dynamic resource provisioning, and cooling [20]. In addition, server-related thermal information, such as inlet/outlet air temperature and air mover speed, is utilized to create thermal and power maps with the ultimate goal to monitor the real-time status of a DC [21].

A majority of previously listed research work focuses on simulations or numerical modeling [2,16–20] as well as on empirical studies involving R&D or small-scale data centers [16,21].

Thus, there is a need for more empirical research involving real thermal-related data for large scale data centers. Undeniably, it is tremendously beneficial to identify hotspots and the air dynamics (particularly its negative effects) within a DC IT room. Such useful evidence-based information will help DC operators improve DC thermal management and ensure uninterrupted steady computing system operations. This is made possible when affected servers continue to perform graceful degradation-related computations or enter the 'switch off' mode once the temperature threshold is breached. Thermal management could be improved in a number of ways based on evidence-based analysis. For example, some corrective actions could be: identify cold air leakages and erect isolating panes; adjust speed, volume, and direction of the cold air stream; apply free cooling wherever possible or adjust the humidity levels. An exhaustive guideline for DC thermal management improvement can be found in [7].

A crucial step forward in DC thermal management related research could be adherence to the recommended thermal management framework [22] at varying DC granularity levels. As a part of the framework, thermal metrics have been created by research and enterprise DC communities [10]. Employment of the metrics aims to reveal the underlying causes of thermal-related issues within a DC IT room and to assess the overall thermal conditions of the room. A recently proposed holistic DC assessment method is based on biomimicry [23]. This integrates data on energy consumption for powering and cooling ICT equipment.

This paper is an extension of the previous authors' work [10,11,24–28], which focus on real DC thermal monitoring data. In detail, this current research focuses on the analysis of DC IT room thermal characteristics to uncover ways to render a more effective cooling system as well as explore possibilities to employ machine learning techniques to address this issue. Appropriate data analytics techniques have been applied on real server-level sensor data to identify potential risks caused by the possible existence of negative covert physical processes related to the cooling strategy [2]. In summary, this work is based on the analysis of DC thermal characteristics using machine learning (ML) techniques. ML has been generally employed for virtual machines allocation, global infrastructure management, prediction of electricity consumption, and availability of renewable energy [29]. Thus far, there is work on ML for thermal characteristics assessment and weather conditions prediction, but only limited available work on thermal management. Typically, Computational Fluid Dynamics (CFD) techniques have been employed for the exploration of DC thermal management. Their drawbacks are high computational power and memory requirements. Therefore, the added value of this research is an evidence-based recommendation for a cooling system for more targeted temperature management through thermal characteristics analysis for localization of overheated areas in the DC IT room.

3. Methodology

This section discusses the thermal characteristics analysis of an ENEA R.C. Portici cluster CRESCO 6. An ML clustering technique was chosen for a more in-depth analysis of overheated servers' localization based on an available dataset of CRESCO6 server temperature measurements. The terms "server" and "node" are used interchangeably in this work, while "hotspot" is utilized to indicate an overheated area next to a server in the IT room and results in an overheated or undercooled server. The drawback of a typical analysis of temperature measurements is that it could not locate the specific nodes which cause rack overheating. Hence, to address this issue, we have applied node clustering to localize potentially harmful hotspots. To identify overheated areas in the CRESCO6 group of nodes, we sequentially grouped the nodes into clusters characterized by higher or lower surrounding air temperature [9]. The term "group of nodes" stands for the DC "cluster" (note that this term is not used to avoid its confusion with the term "cluster", which is the outcome of running an ML clustering algorithm).

3.1. Cluster and Dataset Description

Thermal analysis was based on monitoring data of the CRESCO6 cluster in the premises of ENEA-Portici Research Center (R.C.). Data collected were cluster power consumption of IT equipment (servers) and measurements of ambient air temperature. This cluster has been up and running since May 2018. It is used to augment the computing resources of the CRESCO4 system, already installed and still operating in the Portici Research Center. The reason for the augmentation is due to the rise in demand for analytic activities. Thus, with the addition of the cluster CRESCO6, the overall computing capability of ENEA R.C. has increased up to seven-fold. The cluster comprises 418 Lenovo nodes housed in a total of 5 racks. Each node includes two CPUs, each with 24 cores (with a total of 20,064 cores). This pool of resources is aimed to support Research and Development activities in ENEA Research Center [9].

Measurements of ambient air temperature were recorded during two phases. The first phase was during cluster set up and performance tuning (subject to thermal control strategies) and other indicator specifications during the months of May–July 2018. Subsequently, end-users were allowed to submit and run their jobs and relevant parameters had been monitored and measured for approximately 9 months (September 2018–February 2019). The measurements were paused in August 2018 as shown in Figure 1 [9]. Data collected encompassed all 216 nodes, out of which 214–215 nodes were consistently monitored, and the other 1–2 nodes had missing values or were turned off. The monitoring system consisted of an energy meter, a power meter of CPU, RAM and computing power utilization of every node, and CPU temperature for both processing units of each node with thermal sensors installed inside the servers, at the inlet and exhaust air locations in cold and hot aisles respectively (i.e., placed in the front and rear parts of every node).

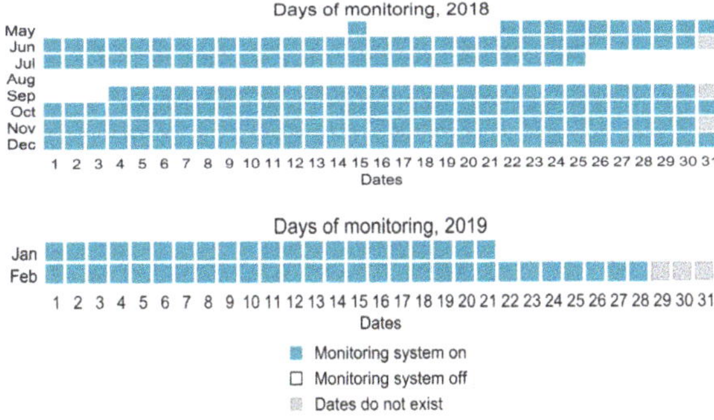

Figure 1. Period of available measurements data in May–December 2018 and January–February 2019.

3.2. Data Analytics

Variation of the air temperature was captured and analyzed for different areas of the IT room. The variability of thermal data and uncertainty in defining temperature thresholds for overheated areas has provided a justification for the use of an unsupervised learning technique. Hence, a k-means algorithm was employed to address the limitations of typical statistical techniques and cluster the servers according to their surrounding air temperature. Silhouette metric and within-cluster sum of squares were used to first determine the number of clusters. Available thermal characteristics (i.e., exhaust air temperature, readings of CPU temperature) served as inputs to the k-means algorithm. A set of surrounding air temperature measurements for the nodes was clustered the same number of times as the number of measurements taken for the batch of all the nodes. Subsequently, the resulting series of cluster labels were intersected to unravel nodes (distinguishable by their IDs) that frequently occurred in the high-temperature cluster.

An adapted data lifecycle methodology was employed for this work, as depicted in Figure 2. The methodology comprises several data preprocessing steps, data analysis, followed by interpretation of the results and their exploitation in the form of recommendations for the DC under consideration [9]. A detailed discussion of all data analytics stages represented in Figure 2 are found in the ensuing section.

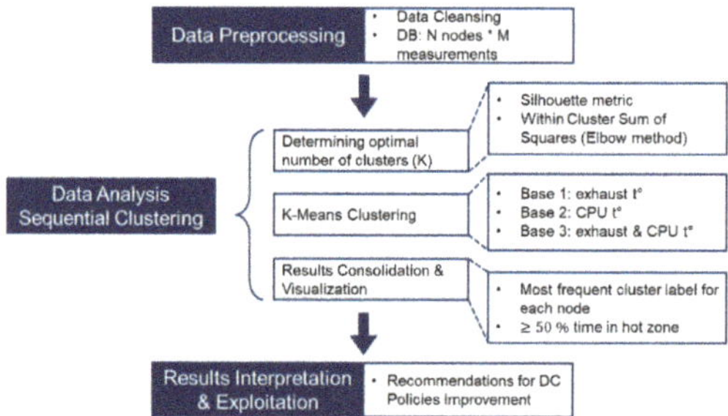

Figure 2. Data analysis lifecycle methodology adapted to sequential clustering of DC servers based on their thermal characteristics.

The data preprocessing step consisted of data cleansing of zero and missing values and formatting. The dataset was organized as shown in Table 1. This table summarizes the results of monitoring of the overall number of nodes in a computing cluster, N. In addition, data preprocessing involved timestamps formatting for further exploitation. In detail, the system was configured so that the monitoring system recorded the thermal data for every node with an interval of around 15 min, including a slight latency between each pair of consecutive readings of temperature sensors around the nodes. The readings resulted in a set of N rows with the information for every node ID.

Table 1. Dataset using for clustering analysis.

Time Label	Real Time of Measurement	Node ID	Inlet T (°C)	Exhaust T (°C)	CPU 1 T (°C)	CPU 2 T (°C)	Cluster Label
t_1	$t_1 + t_{1n_1}$ $t_1 + t_{1n_N}$	n_1 n_N	$T_{in_{11}}$ $T_{in_{1N}}$	$T_{exh_{11}}$ $T_{exh_{1N}}$	$T_{CPU1_{11}}$ $T_{CPU1_{1N}}$	$T_{CPU2_{11}}$ $T_{CPU2s_{1N}}$	$C_1\ base_{range}$ $C_1\ base_{range}$
t_2	$t_2 + t_{2n_1}$ $t_2 + t_{2n_N}$	n_1 n_N	$T_{in_{21}}$ $T_{in_{2N}}$	$T_{exh_{21}}$ $T_{exh_{2N}}$	$T_{CPU1_{21}}$ $T_{CPU1_{2N}}$	$T_{CPU2_{21}}$ $T_{CPU2s_{2N}}$	$C_2\ base_{range}$ $C_2\ base_{range}$

The data analysis step included several substages. The sequential clustering substage encompassed the investigation into the optimal number of clusters followed by server clustering into groups (with three possible levels: low, medium, and high) of the surrounding air temperature. The results were further consolidated to ascribe final cluster labels for each server (i.e., low, medium, or high temperature) based on the frequency of occurrences for each node label in the sequence of results [9]. Clustering was performed M times, where M is the overall number of time labels at which measurements were taken for all cluster nodes. Each new set of monitoring system readings was labeled with a time label $t_1, \ldots, t_M$. The exact timestamp for the extracted information was marked with $t_i + t_{in_j}$ for every node j. Depending on the available dataset, a number of relevant features described the thermal state of every node and their different combinations could be used as a basis for clustering (RO.2 will be more considered in detail in Section 4). Thus, we introduce *base* in the last column of Table 1 which denotes the basis for clustering, i.e., a combination of temperature measurements taken as k-means input (see Results and Discussions for details). The indicator *base* also corresponds to the temperature of the cluster centroid [9].

In this work, k-means was chosen as a clustering algorithm due to the following reasons (RO.1):

1. The number of features used for clustering was small. Therefore, the formulated clustering problem was simple and did not require complex algorithms;
2. K-means has linear computational complexity and is fast to use for the problem in question. While the formulation of the problem is simple, it requires several thousands of repetitions of clustering for each set of N nodes. From this point of view, the speed of the algorithm becomes an influential factor;
3. K-means has a weak point, namely the random choice of initial centroids, which could lead to different results when different random generators are used. This does not pose any issue in this use case since the nodes are clustered several times based on sets of measurements taken at different timestamps and minor differences brought by the randomness are mitigated by the repetition of the clustering procedure.

The number of clusters, K, is an unknown parameter that also defines the number of ranges for $C_{i\ base_{range}}$. It is estimated for each of the three combinations or *bases* using two metrics separately: the average silhouette coefficient and within cluster sum of squares (WCSS) metric [9,30,31] (RO.3). The application of these two indices to derive the suitable number of thermal ranges or clusters is shown in Appendix A. In brief, the silhouette coefficient was computed for each clustered sample of size N and showed the degree of isolation for the clusters, thus, indicating the quality of clustering. The +1 value of silhouette index for a specific number of clusters, K, indicated the high density of clusters, −1 showed incorrect clustering, and 0 stood for overlapping clusters. Therefore, we focused on local maxima of this coefficient. WCSS was used in the Elbow method of determining the number of clusters and was used here to support the decision obtained from the silhouette coefficient estimation. It measured the compactness of clusters, and the optimal value of K was the one that resulted in the "turning point" or the "elbow" of the WCSS (K) graph. In other words, if we increase the number of clusters after reaching the elbow point, it does not result in significant improvement of clusters compactness. Although it could be argued that other features could be additionally used for determining the number of clusters, the combination of the two aforementioned methods had converged on the same values of K for our chosen *bases*, which was assumed to be sufficient for this current research.

Once we obtained the optimal parameter K, we performed the clustering procedure for the chosen *bases*. For a fixed *base*, the sequence of cluster labels was examined for every node. Based on the frequency of occurrences for each cluster label (low, medium, or hot air temperature labels), the node was ascribed the final most frequently occurring cluster label $C_{base_{range}}$ in the sequence and was assigned to the corresponding set of nodes as $N_{base_{range}}$. Furthermore, we took the intersection of the sets of nodes in the hot air range for every *base*. Thus, we obtained the IDs of the nodes that were most frequently labeled as the nodes in "danger" or hot areas of the IT rack by three clustering algorithms: $N_{hot} = \cap_{bases}\{N_{base_{hot\ range}}\}$ (RO.4). In the following section, we will discuss the results of k-means sequential clustering and identify the nodes that occurred in the overheated areas more frequently than others.

4. Results and Discussions

For every combination of measured thermal data, the results of servers clustering into cold, medium, and hot temperature ranges had been further analyzed to calculate the frequency of occurrences of each node in each cluster and determined their final frequency label (i.e., cluster label or temperature range). These labels were further intersected with labels obtained for different bases. Each set of $N = 216$ nodes was clustered at once, followed by temperature-based clustering for the same set of nodes using measurements taken at the next timestamp. This process is referred to here as sequential clustering. The indicator *base* of the input temperature data was used in three possible combinations of available thermal data: exhaust air (*base* 1), CPU (*base* 2), and exhaust air and CPU temperature measurements (*base* 3) (RO.2).

The dataset contained $M = 15,569$ sets of temperature measurements. Each M sets consisted of 216 node-level temperature measurements of the sensor data. The sensors were installed in different locations with respect to the node: in the front (inlet), rear (exhaust) of every node, and two sensors inside each node (CPU temperature). The optimal number of clusters K was influenced by the *base* chosen for clustering. Using the silhouette metric and WCSS, we obtained the optimal number of K for *bases* 1–3 (exhaust, CPU, and exhaust and CPU measurements) and the K values equaled to three, five, and three clusters, respectively [9].

During sequential clustering, each node was labeled with a particular temperature range cluster. Since clustering was repeated for each set of measurements grouped by time label, every node was repeatedly clustered several times and tagged with different labels while the algorithm was in progress (RO.4). Figure 3a–c shows the results of sequential clustering, where one set of three or five vertically aligned graphs represents the result for sequential clustering using one input *base*. In detail, Figure 2 compares how frequently each of the 216 nodes is labeled with low, medium, or hot temperature range. Every vertical graph corresponds to the proportion of occurrences (in %) of low, medium, or hot temperature labels for one node. This figure indirectly implies the incidence rate, "duration", or tendency of a particular node experiencing a certain temperature range (see legend in Figure 3a–c). Here, the medium temperature range label most frequently occurred for the majority of the nodes for all cluster *bases*. We also observe that some nodes remain in the hot range for more than 50% of clustering iterations. This information could alternatively be obtained if the temperature levels (lower and upper bounds of low, medium, and hot temperature ranges) were preset by an expert, but for this research, this estimation is not available. Therefore, the consideration of the clustering results is crucial for the DC in a situation when temperature ranges are unavailable. Figure 2 provides a means for asynchronous assessment of the thermal state of the IT room and unravels relevant thermal trends.

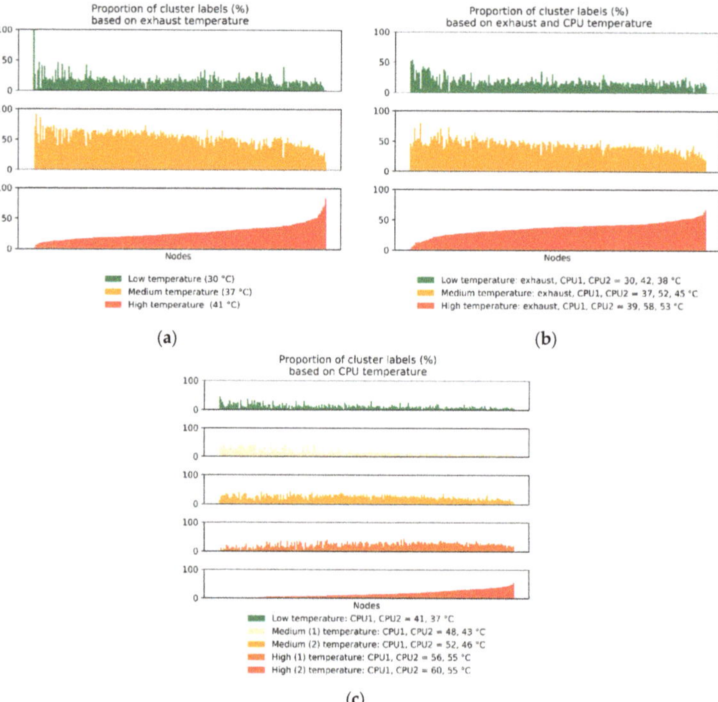

Figure 3. The ratio of nodes labeled by different temperature ranges (low, medium, and hot) based on different thermal data input (*base*): (**a**) Exhaust temperature, (**b**) Exhaust and CPU temperature, (**c**) CPU temperature. The legend includes coordinates of cluster centroids.

In cases where nodes remain in the hot range for a prolonged period or frequently fall in this range, it implies that they are overheated (and the cooling appears to be ineffective). Consequently, this could cause hardware degradation where the nodes have reduced reliability and accessibility as they automatically switch to lower power mode when overheated. Therefore, we continued with the analysis to identify the actual node IDs that had most frequently been clustered within the hot air temperature ranges. Table 2 provides an insight into the ratio of nodes with the highest frequency of occurrences in cold, medium, or hot air temperature range (RO.5). Depending on the cluster base, 50% to 86% of all nodes had the highest frequency of occurrence in the medium range. The hot range encompassed 11%–37% of all nodes, and only 0.5%–4% had been clustered within the cold range.

Table 2. Ratio of cluster sizes and intersection of node labels from three thermal data combinations (*bases*).

Cluster Type Ratio%	$N_{exh_{cold}}$ 2.8	$N_{exh_{med}}$ 86.0		$N_{exh_{hot}}$ 11.2	
Cluster Type Ratio%	$N_{CPU_{cold}}$ 4.2	$N_{CPU_{med1}}$ 20.0	$N_{CPU_{med2}}$ 28.4	$N_{exh_{hot1}}$ 31.2	$N_{exh_{hot2}}$ 16.2
Cluster Type Ratio%	$N_{exh\_CPU_{cold}}$ 2.0	$N_{exh\_CPU_{med}}$ 63.0		$N_{exh\_CPU_{hot}}$ 35.0	
Cluster Type Ratio%	N_{cold} 0.5	N_{med} 40.0		N_{hot} 8.0	
Hot Range Node ID	-	-		30, 31, 32, 45, 46, 48, 68, 79, 94, 96, 105, 117, 118, 120, 182, 183, 189, 198	

Finally, the results of sequential clustering with three bases were intersected for cross-validation purposes. These results in the intersection of nodes were labeled as cold, medium, and hot surrounding air range. One node (equal to 0.5% of nodes) was labeled as the cold air temperature range for all three *bases*. The cluster characterized by the medium air temperature range had the largest intersection among the *bases* (i.e., more than 90%), while 8% (or 18 nodes) were binned in the hot air temperature range most frequently using all three *bases*. The highlight of this exploration is that we were able to identify the IDs of the most frequently overheated nodes. DC operators could further exploit this evidence-based information to improve thermal conditions in the cluster IT room. Possible corrective actions to mitigate overheating of the localized nodes could be to improve the existing natural convection cooling by directing the cold air to the hottest nodes. In addition, DC operators could update the load scheduling and decrease the workload of the identified nodes to indirectly prevent overheating (RO.6). In summary, DC operators could improve resource allocation policies and cooling strategies to effectively address this issue.

This current paper has contributed to thermal characteristics awareness for a real DC cluster and addressed the issue of servers overheating. This has two positive effects in terms of sustainability. Firstly, local overheating could be considered as an IT room thermal design pitfall. It leads to a high risk of hardware degradation for servers that are frequently and/or for long time exposed to high surrounding air temperature. From this perspective, the localization of hot regions of the IT room performed in this study (via a ML technique) is crucial for providing a better overview of thermal distribution around the servers which could be fed into better thermal control and management strategies. In other words, future thermal management improvements could be aligned to the direction provided in this study with the aim of mitigating the abovementioned risk. Secondly, a clustering technique used in this phase requires less computational resources compared to heatmaps, computational fluid dynamics modeling and/or simulations performed on existing simulation packages [9]. Therefore, this work evidences the benefit of less computationally intensive analytical techniques (in yielding sufficient information) to incentivize improvement of thermal conditions (through even thermal distribution) in data centers. In summary, conclusions that could be drawn from this research are that a majority of the nodes were located in medium and hot air temperature ranges. Joint results of three clustering algorithms had shown that that 8% of cluster servers were most frequently characterized as having hot surrounding air temperature. Based on this evidence, we have formulated a list of recommendations (see subsequent section) to address the problem of repeated or prolonged overheating of servers (RO.6).

5. Conclusions and Future Work

Analysis of IT and cooling systems is necessary for the investigation of DC operations-related energy efficiency. A reliable cooling system is essential to produce a persistent flow of cold air to cool servers that could be overheated due to an increasing demand in computation-intensive applications. To reiterate, Patterson [8] has maintained there is an impact of DC ambient temperature on energy consumption of IT equipment and systems. However, in this paper, the focus is on thermal characteristics analysis of an IT room. The research methodology discussed in this paper includes statistical analysis of IT room thermal characteristics and the identification of individual servers that frequently occur in the overheated areas of the IT room (using a machine learning algorithm). Clustering techniques are used for hotspot localization as well as categorization of nodes (i.e., servers) based on surrounding air temperature ranges. This methodology has been applied to an available dataset with thermal characteristics of an ENEA Portici CRESCO6 computing cluster. In summary, this paper has presented a proposed methodology for IT room thermal characteristics assessment of an air-cooled DC cluster located in a geographical region where free air cooling is unavailable. The steps involved for evidence-based targeted temperature management (to be recommended for air-cooled DCs) are as follows:

1. Explore the effectiveness of the cooling system by firstly uncovering nodes with hot range IDs (e.g., change direction, volume, speed of cooling air). Additionally, directional cooling could be recommended (e.g., spot cooling to cool overheated nodes). Next, unravel covert factors that lead to nodes' repetitive overheating (e.g., location next to the PDUs that have higher allowable temperature ranges);
2. Revise cluster load scheduling so that these frequently overheated servers are not overloaded in the future (note: this is to enable an even thermal distribution within the IT room. See [11] for details). In other words, it is recommended to formulate a resource allocation policy for the purpose of a more even thermal distribution of ambient air temperature;
3. Perform continuous environmental monitoring of the IT room and evaluate the effectiveness of recommended actions and their influence on the ambient temperature.

To reiterate, the approaches covered in this work are transferrable for thermal characteristics analysis in any air-cooled DC context enabled with a thermal monitoring system. This study illustrates the applicability of the best practices and guidelines to a real DC and uses an ML approach to perform IT room thermal characteristics assessment. This work could be extended by incorporating an integrated thermal management with existing energy efficiency policies-related research (e.g., energy awareness [15]; job scheduling using AI [32], temporal-based job scheduling [33], work-load aware scheduling [34], and queue theory [35]; resource utilization [36] of multiple applications using annealing and particle swarm optimization [37]). Another direction that could be taken could be the energy efficiency policies and waste heat utilization [26].

Author Contributions: Conceptualization A.G., M.C., A.-L.K., E.R. and J.-P.G.; methodology, A.G., M.C., A.-L.K.; data analysis, A.G.; data curation, M.C.; writing—original draft preparation, A.G., M.C., A.-L.K. and J.-P.G.; writing—review and editing, A.G., M.C., A.-L.K.; project administration, E.R. All authors have read and agreed to the published version of the manuscript.

Funding: This research was funded by EMJMD PERCCOM Project [38].

Acknowledgments: This work was supported by EMJMD PERCCOM Project [38]. During the work on the project, the author A.G. was affiliated with ENEA-R.C. Casaccia, Rome, Italy. Moreover, the authors extend their gratitude to the research HPC group at the ENEA-R.C. Portici for the discussion of ENEA-Data Center control and modeling.

Conflicts of Interest: The authors declare no conflict of interest.

Appendix A

Several approaches are widely used by data scientists to identify the optimal number of clusters. However, it is worth noting that none of the approaches is considered accurate, instead they provide a suggestion for the number of clusters. The current study applies two indices: WCSS, also known as an elbow method, and average Silhouette Index [32]. We apply a systematic approach and calculate these metrics for different values of cluster numbers, *K*. Subsequently, an optimal value is chosen based on the values of the indices. WCSS characterizes the compactness of the cluster and is defined as follows:

$$WCSS(K) = \sum_{j=1}^{K} \sum_{x \in C_j} \|x - \mu_j\|^2, \quad (1)$$

where K is the number of clusters, C is a set of clusters ($C_1, C_2, \ldots C_j$), and μ_j represents a certain cluster sample mean. The target value of this metric should be a possible minimized value. Practically, this refers to a point where the value of the metric continues to decrease but at a significantly slower rate in comparison to a smaller number of clusters and is considered an optimal one. It is visually associated with an "elbow" of the graph. The justification for choosing an elbow point is that with the increase in the number of clusters, the metric decreases only slightly, while computations become increasingly more intensive.

The second method applied in this paper is the average silhouette. It is computed using average silhouette index over all data points (or cluster members). It estimates within-cluster consistency and should be maximized to achieve the effective cluster split. The formula for the Silhouette index is as follows:

$$s(i) = \frac{b(i) - a(i)}{max\{b(i), a(i)\}}, \quad (2)$$

where, a is the mean distance between one cluster member and all other members of the same cluster. Parameter b is the distance between a cluster member and all other points in the nearest cluster. We depict the results of this metrics calculation in Figure A1. It shows the metrics for one-step of sequential clustering based on server exhaust air temperature. At $K = 3$, we observe a local maximum for silhouette index and an elbow point of WCSS graph.

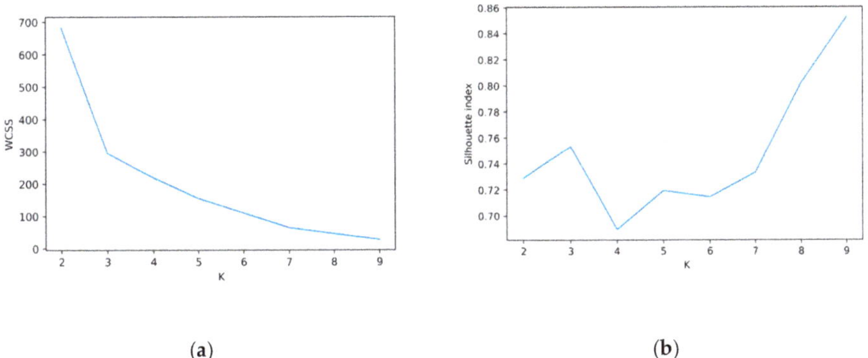

(a) (b)

Figure A1. (**a**) WCSS; (**b**) Average Silhouette Index. Both indices are computed for one step of the sequential clustering procedure based on the exhaust air temperature.

References

1. Hashem, I.A.T.; Chang, V.; Anuar, N.B.; Adewole, K.S.; Yaqoob, I.; Gani, A.; Ahmed, E.; Chiroma, H. The role of big data in smart city. *Int. J. Inf. Manag.* **2016**, *36*, 748–758. [CrossRef]
2. Zhang, K.; Zhang, Y.; Liu, J.; Niu, X. Recent advancements on thermal management and evaluation for data centers. *Appl. Therm. Eng.* **2018**, *142*, 215–231. [CrossRef]
3. Datacenter Knowledge. A Critical Look at Mission-Critical Infrastructure. 2018. Available online: https://www.datacenterknowledge.com/industry-perspectives/critical-look-mission-critical-infrastructure (accessed on 26 June 2020).
4. Hartmann, B.; Farkas, C. Energy efficient data centre infrastructure—Development of a power loss model. *Energy Build.* **2016**, *127*, 692–699. [CrossRef]
5. He, Z.; Ding, T.; Liu, Y.; Li, Z. Analysis of a district heating system using waste heat in a distributed cooling data center. *Appl. Therm. Eng.* **2018**, *141*, 1131–1140. [CrossRef]
6. Nadjahi, C.; Louahlia, H.; Lemasson, S. A review of thermal management and innovative cooling strategies for data center. *Sustain. Comput. Inform. Syst.* **2018**, *19*, 14–28. [CrossRef]
7. AT Committee. Data Center Power Equipment Thermal Guidelines and Best Practices Whitepaper. ASHRAE, Tech. Rep., 2016. Available online: https://tc0909.ashraetcs.org/documents/ASHRAE_TC0909_Power_White_Paper_22_June_2016_REVISED.pdf (accessed on 6 June 2019).
8. Patterson, M.K. The effect of data center temperature on energy efficiency. In Proceedings of the 11th Intersociety Conference on Thermal and Thermomechanical Phenomena in Electronic Systems, Orlando, FL, USA, 28–31 May 2008; pp. 1167–1174. [CrossRef]
9. Grishina, A. Data Center Energy Efficiency Assessment Based on Real Data Analysis. Unpublished PERCCOM Masters Dissertation. 2019.
10. Capozzoli, A.; Serale, G.; Liuzzo, L.; Chinnici, M. Thermal metrics for data centers: A critical review. *Energy Procedia* **2014**, *62*, 391–400. [CrossRef]
11. De Chiara, D.; Chinnici, M.; Kor, A.-L. Data mining for big dataset-related thermal analysis of high performance (HPC) data center. In *International Conference on Computational Science*; Springer: Cham, NY, USA, 2020; pp. 367–381.
12. Chinnici, M.; Capozzoli, A.; Serale, G. Measuring energy efficiency in data centers. In *Pervasive Computing: Next Generation Platforms for Intelligent Data Collection*; Dobre, C., Xhafa, F., Eds.; Morgan Kaufmann: Burlington, MA, USA, 2016; Chapter 10; pp. 299–351. ISBN 9780128037027.
13. Infoworld. Facebook Heat Maps Pinpoint Data Center Trouble Spots. 2012. Available online: https://www.infoworld.com/article/2615039/facebook-heat-maps-pinpoint-data-center-trouble-spots.html (accessed on 20 June 2020).
14. Bash, C.E.; Patel, C.D.; Sharma, R. Efficient thermal management of data centers—Immediate and long-term research needs. *HVAC&R Res.* **2003**, *9*, 137–152. [CrossRef]
15. Fernández-Cerero, D.; Fernández-Montes, A.; Velasco, F.P. Productive Efficiency of Energy-Aware Data Centers. *Energies* **2018**, *11*, 2053. [CrossRef]
16. Fredriksson, S.; Gustafsson, J.; Olsson, D.; Sarkinen, J.; Beresford, A.; Kaufeler, M.; Minde, T.B.; Summers, J. Integrated thermal management of a 150 kW pilot Open Compute Project style data center. In Proceedings of the 2019 IEEE 17th International Conference on Industrial Informatics (INDIN), Helsinki, Finland, 22–25 July 2019; Volume l, pp. 1443–1450.
17. Srinarayana, N.; Fakhim, B.; Behnia, M.; Armfield, S.W. Thermal performance of an air-cooled data center with raised-floor and non-raised-floor configurations. *Heat Transf. Eng.* **2013**, *35*, 384–397. [CrossRef]
18. Schmidt, R.R.; Cruz, E.E.; Iyengar, M. Challenges of data center thermal management. *IBM J. Res. Dev.* **2005**, *49*, 709–723. [CrossRef]
19. MirhoseiniNejad, S.; Moazamigoodarzi, H.; Badawy, G.; Down, D.G. Joint data center cooling and workload management: A thermal-aware approach. *Future Gener. Comput. Syst.* **2020**, *104*, 174–186. [CrossRef]
20. Fang, Q.; Wang, J.; Gong, Q.; Song, M.-X. Thermal-aware energy management of an HPC data center via two-time-scale control. *IEEE Trans. Ind. Inform.* **2017**, *13*, 2260–2269. [CrossRef]
21. Zhang, S.; Zhou, T.; Ahuja, N.; Refai-Ahmed, G.; Zhu, Y.; Chen, G.; Wang, Z.; Song, W.; Ahuja, N. Real time thermal management controller for data center. In Proceedings of the Fourteenth Intersociety Conference on

Thermal and Thermomechanical Phenomena in Electronic Systems (ITherm), Orlando, FL, USA, 27 May 2014; pp. 1346–1353.

22. Sharma, R.; Bash, C.; Patel, C.; Friedrich, R.; Chase, J.S. Balance of Power: Dynamic Thermal Management for Internet Data Centers. *IEEE Internet Comput.* **2005**, *9*, 42–49. [CrossRef]
23. Kubler, S.; Rondeau, E.; Georges, J.P.; Mutua, P.L.; Chinnici, M. Benefit-cost model for comparing data center performance from a biomimicry perspective. *J. Clean. Prod.* **2019**, *231*, 817–834. [CrossRef]
24. Capozzoli, A.; Chinnici, M.; Perino, M.; Serale, G. Review on performance metrics for energy efficiency in data center: The role of thermal management. *Lect. Notes Comput. Sci.* **2015**, *8945*, 135–151.
25. Grishina, A.; Chinnici, M.; De Chiara, D.; Guarnieri, G.; Kor, A.-L.; Rondeau, E.; Georges, J.-P. DC Energy Data Measurement and Analysis for Productivity and Waste Energy Assessment. In Proceedings of the 2018 IEEE International Conference on Computational Science and Engineering (CSE), Bucharest, Romania, 29–31 October 2018; IEEE: Piscataway, NJ, USA, 2018; pp. 1–11, ISBN 978-1-5386-7649-3.
26. Koronen, C.; Åhman, M.; Nilsson, L.J. Data centres in future European energy systems—Energy efficiency, integration and policy. *Energy Effic.* **2019**, *13*, 129–144. [CrossRef]
27. Grishina, A.; Chinnici, M.; De Chiara, D.; Rondeau, E.; Kor, A.L. *Energy-Oriented Analysis of HPC Cluster Queues: Emerging Metrics for Sustainable Data Center*; Springer: Dubrovnik, Croatia, 2019; pp. 286–300.
28. Grishina, A.; Chinnici, M.; Kor, A.L.; Rondeau, E.; Georges, J.P.; De Chiara, D. Data center for smart cities: Energy and sustainability issue. In *Big Data Platforms and Applications—Case Studies, Methods, Techniques, and Performance Evaluation*; Pop, F., Ed.; Springer: Berlin, Germany, 2020.
29. Athavale, J.; Yoda, M.; Joshi, Y.K. Comparison of data driven modeling approaches for temperature prediction in data centers. *Int. J. Heat Mass Transf.* **2019**, *135*, 1039–1052. [CrossRef]
30. Kaufman, L.; Rousseeuw, P.J. *Finding Groups in Data: An Introduction to Cluster Analysis*; John Wiley & Sons: Hoboken, NJ, USA, 2009.
31. Kassambara, A. (Ed.) Determining the Optimal Number of Clusters: 3 Must Know Methods. Available online: https://www.datanovia.com/en/lessons/determining-the-optimal-number-of-clusters-3-must-know-methods/ (accessed on 6 May 2019).
32. Fernández-Cerero, D.; Fernández-Montes, A.; Ortega, J.A. Energy policies for data-center monolithic schedulers. *Expert Syst. Appl.* **2018**, *110*, 170–181. [CrossRef]
33. Yuan, H.; Bi, J.; Tan, W.; Zhou, M.; Li, B.H.; Li, J. TTSA: An Effective Scheduling Approach for Delay Bounded Tasks in Hybrid Clouds. *IEEE Trans. Cybern.* **2017**, *47*, 3658–3668. [CrossRef]
34. Yuan, H.; Bi, J.; Zhou, M.; Sedraoui, K. WARM: Workload-Aware Multi-Application Task Scheduling for Revenue Maximization in SDN-Based Cloud Data Center. *IEEE Access* **2018**, *6*, 645–657. [CrossRef]
35. Fernández-Cerero, D.; Irizo, F.J.O.; Fernández-Montes, A.; Velasco, F.P. Bullfighting extreme scenarios in efficient hyper-scale cluster computing. In *Cluster Computing*; Springer: Berlin/Heidelberg, Germany, 2020; pp. 1–17. [CrossRef]
36. Fernández-Cerero, D.; Fernández-Montes, A.; Jakobik, A.; Kołodziej, J.; Toro, M. SCORE: Simulator for cloud optimization of resources and energy consumption. *Simul. Model. Pract. Theory* **2018**, *82*, 160–173. [CrossRef]
37. Bi, J.; Yuan, H.; Tan, W.; Zhou, M.; Fan, Y.; Zhang, J.; Li, J. Application-Aware Dynamic Fine-Grained Resource Provisioning in a Virtualized Cloud Data Center. *IEEE Trans. Autom. Sci. Eng.* **2015**, *14*, 1172–1184. [CrossRef]
38. Klimova, A.; Rondeau, E.; Andersson, K.; Porras, J.; Rybin, A.; Zaslavsky, A. An international Master's program in green ICT as a contribution to sustainable development. *J. Clean. Prod.* **2016**, *135*, 223–239. [CrossRef]

Article

Stochastic Modeling of the Levelized Cost of Electricity for Solar PV

Chul-Yong Lee [1] and Jaekyun Ahn [2,*]

[1] School of Business, Pusan National University, 2, Busandaehak-ro 63beon-gil, Geumjeong-gu, Busan 46241, Korea; cylee7@pusan.ac.kr

[2] Korea Energy Economics Institute, 405-11, Jongga-ro, Jung-Gu, Ulsan 44543, Korea

* Correspondence: jkahn@keei.re.kr; Tel.: +82-52-714-2265

Received: 25 April 2020; Accepted: 10 June 2020; Published: 11 June 2020

Abstract: With the development of renewable energy, a key measure for reducing greenhouse gas emissions, interest in the levelized cost of electricity (LCOE) is increasing. Although the input variables used in the LCOE calculation, such as capacity factor, capital expenditure, annual power plant operations and maintenance cost, discount and interest rate, and economic life, vary according to region and project, most existing studies estimate the LCOE by using a deterministic methodology. In this study, the stochastic approach was used to estimate the LCOE for solar photovoltaic (PV) in South Korea. In addition, this study contributed to deriving realistic analysis results by securing the actual data generated in the solar PV project compared to the existing studies. The results indicate that the LCOE for commercial solar power ranged from KRW 115 (10 cents)/kWh to KRW 197.4 (18 cents)/kWh at a confidence level of 95%. The median was estimated at KRW 160.03 (15 cents)/kWh. The LCOE for residential solar power ranged from KRW 109.7 (10 cents)/kWh to KRW 194.1 (18 cents)/kWh at a 95% confidence level and a median value of KRW 160.03 (15 cents)/kWh. A sensitivity analysis shows that capital expenditure has the most significant impact on the LCOE for solar power, followed by the discount rate and corporate tax. This study proposes that policymakers implement energy policies to reduce solar PV hardware and soft costs.

Keywords: LCOE; stochastic; solar PV; South Korea; renewable energy

---

1. Introduction

Since the Paris Agreement came into effect in November 2016, the issue of reducing greenhouse gas (GHG) emissions gained traction globally. As a result, most countries are required to submit and implement Nationally Determined Contributions in an effort to address this issue. For example, South Korea is expected to reduce GHG emissions by 37% from business-as-usual levels by 2030. However, the reduction targets submitted per country are currently lower than the global target of maintaining a temperature rise below 2 °C above pre-industrial levels for this century [1]. Further GHG reduction targets for each country are to be discussed in order to meet the universal target, which is therefore likely to become a major constraint on the global economy.

A transition to renewable energy is one of the key measures for reducing GHG emissions. Solar photovoltaic (PV) is the fastest-growing source of numerous renewable energy sources, leading to a sharp reduction in cost and an increase in demand. Therefore, it is essential to accurately estimate the cost of solar PV and to compare it with other energy sources. To do so, it is necessary to compare the costs incurred for producing equivalent amounts of power. For this reason, many studies have introduced the levelized cost of electricity (LCOE) [2–6].

This study has marginal contributions to the previous study from three perspectives. First, this study considers more sophisticated input variables than previous studies. Most existing studies consider capacity factors, capital expenditure (CAPEX), annual power plant operations and maintenance (O&M),

discount rate, and economic life as input variables [2–5]. In this study, a more realistic analysis is attempted in additional consideration of the project's corporate tax, debt cost, inflation rate, and loan interest rate. Second, existing LCOE-related studies were analyzed from a deterministic point of view. Since the input variables used in the LCOE calculation vary according to region and project, simulation techniques are useful to account for these changes. This study aims to stochastically estimate the LCOE based on a Monte Carlo simulation to consider the variation of input variables. Third, while existing stochastic approach studies subjectively assume input variables, this study derives the optimal distribution using actual data in the case of capacity factor. The distribution is analyzed by using the actual generation data of the solar PV project and the Kolmogorov–Smirnov statistics test.

The target of the analysis is solar PV in South Korea, which is growing rapidly. As a result of stochastic approach using Monte Carlo simulation, significant statistical values such as a reference value, an average, a median value, a standard deviation, a minimum value, and a maximum value are derived. The methodology proposed in this study can be applied to various energy sources in multiple countries globally. In addition, these results are expected to prove valuable in countries' energy policy development and economic analysis.

The structure of this paper is as follows. Section 2 demonstrates the current status of the global solar LCOE. Section 3 introduces the methodology used in this study. Section 4 examines the stochastic LCOE results, and Section 5 discusses the results and presents policy implications.

2. Literature Review

Bhandari and Stadler [7] compared the average LCOE of residential and commercial solar PV in Cologne, Germany, with the electricity rate to determine grid parity. By comparing the LCOE to the high electricity bills of consumers, it was estimated that grid parity would be achieved in 2013 or 2014. However, by considering low wholesale electricity prices, it was found that grid parity would be achieved in 2023. Branker et al. [8] estimated the LCOE by focusing on cases in North America and conducted a sensitivity analysis on the major variables. The LCOE variables of initial investment (installation) costs, investment methods, economic life, and debt redemption period responded sensitively. With developments in financing techniques and industry and technology improvements, it was revealed that solar PV could be more cost effective than traditional energy sources, thereby reaching grid parity more efficiently. Mendicino et al. [4] suggested an appropriate contract price for the Corporate Power Purchase Agreement (CPPA) using the LCOE. CPPA is a contract between electricity consumer and a power generator with renewable energy. The results show that the appropriate price range is between EUR 75/MWh and EUR 100/MWh.

Rhodes et al. [9] calculated the LCOE for 12 plant technologies by county in the United States. For some technologies, the average cost has increased when internalizing the cost of carbon and air pollutants. Including the cost of USD 62/tCO_2 for CO_2 emissions, combined cycle gas turbine, wind and nuclear power showed the lowest LCOE. Clauser and Ewert [10] analyzed the LCOE of geothermal energy and other primary energy. The LCOE was calculated by varying the conditions of geothermal energy, and as a result of these cost comparisons, it was concluded that geothermal energy could be converted into electrical energy at an attractive cost, particularly for steam use in natural or engineered geological reservoirs.

Chadee and Clarke [11] conducted a technical and economic assessment to determine the level of LCOE of wind power in the Caribbean islands. The assessment was conducted on two sites with eight wind turbines ranging from 20 to 3050 kW. Mundada et al. [12] calculated the LCOE for a hybrid system of solar PV, batteries and combined heat power (CHP). Sensitivity analysis of these hybrid systems for LCOE was performed on the capital costs of the three energy subsystems, capacity factor of PV and CHP, efficiency of CHP, natural gas rates and fuel consumption of CHP. The results of sensitivity have provided decision makers with a clear guide to distributed generation LCOE with off-grid PV + battery + CHP systems. Nissen and Harfst [13] proposed an 'Energy price adjusted LCOE' that allows for more accurate LCOE calculations in consideration of rising energy prices.

As such, the LCOE methodology is useful for a variety of countries and for various energy sources. However, the existing studies analyzed LCOE from a deterministic point of view, and thus did not reflect much uncertainty about the input variable. Therefore, this study has a marginal contribution compared to previous studies in that it analyzes the LCOE using a stochastic approach. In addition, in assuming the distribution of the input variable, in the case of capacity factor, the distribution was derived using actual data. Lastly, this study is contributing to deriving a more realistic LCOE in that it considers more specific input variables, such as corporate tax, debt cost, inflation rate, and loan interest rate than previous studies.

3. Methodology

3.1. Levelized Cost of Electricity

The LCOE is the average cost per unit of electricity generated by a particular plant. It is calculated by dividing the present value of the total generation cost of the facility by the present value of total power generation. The LCOE allows the evaluation of the costs in relation to the generated amount of power during the economic lifetime of a plant and across the entire energy generation process, including initial construction capital, operations, and maintenance [14,15].

The total cost incurred in the generation of energy comprises of the initial CAPEX and annual O&M costs. More specifically, the initial CAPEX can be separated into hardware and soft costs. Hardware costs refer to equipment and materials, civil engineering, power generation equipment, and annex buildings, while soft costs include design, permits and authorizations, and construction supervision services. O&M costs cover annual operations and maintenance costs of the power plant and financial services fees, such as insurance premiums [16,17].

The LCOE is affected by construction costs, operations and maintenance costs, the lifespan of the power plant, power generation technology, energy efficiency, system degradation rates, inflation and interest rates, and corporate taxes. The formula for calculating the LCOE may be defined as follows [10]:

$$LCOE_t = \frac{CAPEX_t + \sum_{n=1}^{T} \frac{OM_n + FC_n}{(1+r)^n}}{\sum_{n=1}^{T} \frac{(1-d)^n \times CF \times 365(days) \times 24(hours) \times Capacity}{(1+r)^n}} \quad (1)$$

In the above formula, $CAPEX_t$ refers to initial investment (facilities), which include equipment and materials, the construction of structures, grid connection, permits, design, supervision, and inspection at time t. Indirect costs, OM_n, are the O&M costs at time n; FC_n, the finance costs at time n; r, the discount rate; d, the degradation rate; CF, the capacity factor; capacity, the energy generating capacity of the power plant; and T, the operation period of the power plant. that is interest cost due to debt. In this study, interest cost due to debt is considered as finance cost. The capacity factor is the rate at which the power generator operates for one year. For example, the capacity factor of Korea's solar PV is about 14.78%, which means that it produces only 14.78% of power capacity per year. The discount rate is affected by the inflation rate and the interest rate of safe assets in a country. It could also be interpreted that the LCOE represents the recovery of costs disbursed during the lifetime of a generation facility at a discounted rate r in the form of an equal amount paid annually.

3.2. Stochastic Approach

Economic analysis methodologies can be categorized as deterministic and stochastic models depending on the relationship between the input and output variables. The characteristics of the deterministic model are the relations between the input and output variables that are certain, and that the model allows an analytical solution. Contrarily, the model contains three weaknesses. First, it excludes potential future alternatives as it sets long-term variables at fixed values. Second, if all the scenarios with possible variables are combined, the number of cases increases exponentially, hindering the application of the sensitivity analysis. Third, the model does not allow for the reflection of correlation between variables.

The stochastic model does not enable the development of a solution, meaning that a confidence interval must be identified in the results by incorporating the probabilistic characteristics of the input variables using a simulation technique that generates random numbers [18]. The advantages of the stochastic simulation technique are as follows. First, it enables the estimation of a solution for a comparatively difficult mathematical question. Second, for uncertain variables, it allows for the establishment of a correlation between the probability distributions of the variables. However, a disadvantage is that the estimate produced through the stochastic simulation is an approximate value calculated by repeated sampling, thus requiring a statistical interpretation [19]. Among the stochastic simulation techniques, the Monte Carlo simulation is a method that is used universally. Assuming that the input variable is a probabilistic variable, an adequate probability distribution is selected, and a random number that follows the relevant distribution is produced [20–23].

To explain the value of the Monte Carlo simulation, the probability variable X has a probability density function $f_x(x)$ and assumes an arbitrary function $g(x)$. The expected value of $g(x)$ is as follows:

$$E(g(X)) = \int_{x \in X} g(x) f_x(x) dx \tag{2}$$

To estimate the expected value of g(x) as per the above, n number of samples $(x_1, \ldots, x_n)$ are extracted from a distribution of the probability function X, and the average of g(x) is calculated as below:

$$\widetilde{g}_n(X) = \frac{1}{n}\sum_{i=1}^{n} g(x_i) \tag{3}$$

$\widetilde{g}_n(X)$ is the Monte Carlo estimator of $E(g(X))$, which is based on the law of large numbers. In the case where the weak law of large numbers is expressed as the formula below, it could be concluded that when the number (n) of samples is infinite, the average of g(x) based on the sampling can be found at $E(g(X))$ [24].

$$\lim_{n \to 0} P(\left|\widetilde{g}_n(X) - E(g(X))\right| \geq \varepsilon) = 0 \tag{4}$$

As a result, $\widetilde{g}_n(X)$ satisfies the identity below and becomes the unbiased estimator of $E(g(X))$.

$$E(\widetilde{g}_n(X)) = E(\frac{1}{n}\sum_{i=1}^{n} g(x_i)) = \frac{1}{n}\sum_{i=1}^{n} E(g(x_i)) = E(g(X)) \tag{5}$$

If $g(x)$ is given as a complex function, the integral calculation becomes problematic, and so does finding a solution. However, if the Monte Carlo simulation is used, the expected value of the function can be estimated without having to conduct a complex calculation process [25,26].

In the Monte Carlo simulation, the value and size of the extracted sample has an absolute effect on the results. Therefore, the method used to generate the random number that follows the given probability distribution is highly important. Figure 1 shows the analysis procedure of the Monte Carlo simulation [27]. An appropriate distribution can be selected based on available data, on the judgment of a knowledgeable expert, or on a combination of data and judgment. Factors for judgment are discrete or continuous, having bound or not, number of modes, and symmetric or skewed. The size of the extracted sample depends on the estimated standard deviation, the desired margin of error, and the critical value of the normal distribution for significant level [28]. As a general rule of thumb, 10,000 iterations are used [29].

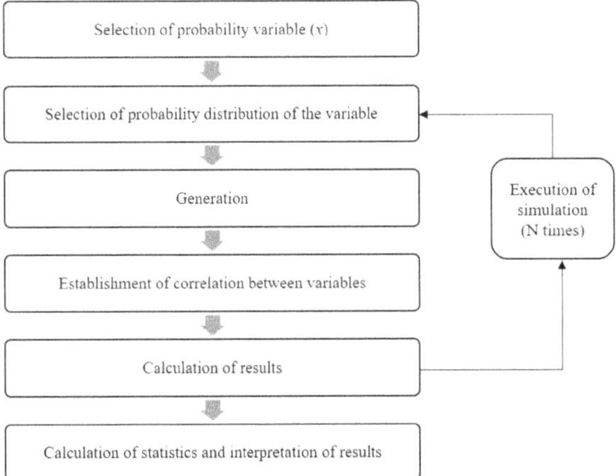

Figure 1. Procedure for the Monte Carlo simulation analysis.

For the purpose of this study, the LCOE analysis of solar PV, the major variables that determine the economics of power generation are the generation amount of power and cost. However, these two variables fluctuate significantly and contain uncertain factors related to the solar radiation amount, technological advancement, and market conditions. When applying the deterministic technique to the economic analysis of photovoltaics, with both volatile and uncertain variables, it becomes difficult to reveal the characteristics of the variables since uncertain future factors are simplified. However, the stochastic simulation method can produce results by considering input variables that are uncertain or volatile through an adequate probability distribution and could therefore be a more appropriate method.

3.3. Sensitivity Analysis

The sensitivity of the contribution of the solar LCOE distribution enables a comparative analysis of the direction and extent of the impact of probability variables on the LCOE. The distribution contribution can be calculated according to the following.

In the first phase, the sample and results of input variables extracted through the simulation are organized in order, and the correlation between the input variable samples and results are identified.

In the second phase and as per Equation (6), the distribution contribution (v_i) of input variable i represents the proportion of the ordinal correlation coefficient squared (R_i^2) over the summation from i to N of the ordinal correlation coefficient squared (R_i^2).

$$v_i = \frac{R_i^2}{\sum_{i}^{N} (R_i^2)} \tag{6}$$

The numerator (ordinal correlation coefficient [R_i]) acquires the original negative (−) or positive (+) sign. The reason being, that when the ordinal correlation coefficient is negative (−), the resulting value decreases as the input variable increases. In contrast, when it is positive (+), the resulting value increases.

4. Empirical Results

4.1. Data

In this section, we apply the above-mentioned stochastic simulation technique to establish the probability distribution for variables with uncertainties. Thereafter, a random sample will be extracted from the relevant distribution, and the resulting value of the probability distribution will be estimated by repeatedly conducting the LCOE calculation.

Table 1 shows the input variables required for the LCOE analysis. The subjects of the analysis are 100 kW commercial facilities installed on the ground and 3 kW residential facilities installed inside buildings. The random variables can be classified as internal and external factors. The internal factors consist of costs and facility characteristics, with the costs comprising CAPEX and O&M costs. In terms of facility characteristics, the capacity factor (which determines the generation amount of power) and system degradation rate are considered random variables. The external factors include the discount rate and corporate tax. Other variables, such as the debt ratio, loan interest rate, inflation, and lifespan, have been granted fixed values, considering their significance and the fact that they fluctuate.

Table 1. Input variables for the solar levelized cost of electricity (LCOE).

	Solar (Commercial)	Solar (Residential)
Standard size	100 kW	3 kW
CAPEX (100 million won/MW)	Normal distribution (average = 16.1, deviation = 10% of average)	Normal distribution (average = 18.3, deviation = 10% of average)
O&M costs (10,000 won/MW·year)	Normal distribution (average = 1167, deviation = 5% of average)	Normal distribution (average = 3737; deviation = 5% of average)
Capacity factor (%)	Logistic distribution (average = 14.78, scale = 0.22)	
Discount rate (%)	Triangular distribution (minimum = 4.5, mode = 5.5, maximum = 7.5)	
Corporate tax (%)	Triangular distribution (minimum = 0, mode and maximum = 24.2)	0
System degradation rate (%)	Triangular distribution (minimum = 0, mode = 0.7, maximum = 0.8)	
Loan interest rate (%/year)	3.46	
Inflation (%)	0.97	
Lifespan (year)	20	
Debt ratio (%)	70	

In the probability distribution, the most appropriate distribution was determined through verification when there were sufficient data samples. Contrarily, the probability distribution cases presented in preceding studies were referred to when the data was lacking or insufficient. A representative study on probability distribution estimation is Spooner [30], which estimated the cost distribution of a construction project and proposed the application methods for normal distribution, log-normal distribution, triangular distribution, beta distribution, and uniform distribution. Uniform distribution is used when there is an insufficient amount of data and the fluctuation range is relatively small. Triangular distribution is used when the maximum likelihood estimation (MLE) is certain, and the information about the maximum and minimum values is considered concrete. However, when the possibility for the reduction of variables is very low, the minimum value does not have to be designated. This study therefore proposed that it is reasonable to estimate the probability distribution for relevant variables using the triangular distributions, which are skewed distributions. This study used the probability simulation analysis software Crystal Ball (version 11).

4.1.1. Capacity Factor

The capacity factor of solar PV is based on data on photovoltaic energy generation provided by the Korea Energy Agency. The goodness-of-fit test was conducted to establish the probability distribution for the capacity factor. The number of capacity factor samples totaled 106,654, and the Kolmogorov–Smirnov (K–S) statistics test is the test method used. As per Equation (7), this test method extracts the maximum value statistics (D_n) by subtracting the cumulative distribution function for the fitting distribution $F(x)$ from the cumulative percentile of the actual measurement data $F_n(x)$. The smaller the statistic, the higher the goodness-of-fit.

$$D_n = \max\left|F_n(x) - F(x)\right| \tag{7}$$

The goodness-of-fit test was conducted with a total of 14 probability distributions, from the logistic distribution to the exponential distribution. Table 2 lists the test results according to each probability distribution. The test results show little difference of the K–S statistics between the logistic distribution and the student t distribution. In this study, the logistic distribution (an average of 14.78%, scale of 0.22%) with the smallest value of the K–S statistics was selected as the probability distribution for capacity factors.

Table 2. Probability distribution test results.

Distribution	K-S Statistics (Dn)	Statistics
Logistic	0.0147	Average = 14.78%, Scale = 0.22%
Student t	0.0149	Intermediate point = 14.78%, Scale = 0.35%, Freedom = 7.28199
Normal	0.0369	Average = 14.78%, Standard deviation = 0.41%
Log-normal	0.0369	Location = −4714.30%, Average = 14.78%, Standard deviation = 0.41%
Beta	0.0376	Minimum = 9.01%, Maximum = 20.54%, Alpha = 100, Beta = 100
Gamma	0.0378	Location = 8.85%, Scale = 0.03%, Form = 207.5021
Weibull	0.0447	Location = 13.02%, Scale = 1.91%, Form = 4.92757
Minimum extreme value	0.0868	Highest probability = 14.98%, Scale = 0.42%
Maximum extreme value	0.1214	Highest possibility = 14.57%, Scale = 0.48%
BetaPERT	0.1801	Minimum = 12.65%, Highest possibility = 14.85%, Maximum = 16.47%
Triangular	0.2268	Minimum = 12.65%, Highest possibility = 14.85%, Maximum = 16.47%
Uniform	0.3409	Minimum = 12.66%, Minimum = 16.46%
Pareto	0.4606	Location = 12.66%, Form = 6.47827
Exponential	0.5933	Ratio = 676.83%

The probability density function (PDF) of a logistic distribution is presented as follows [31]:

$$f(x) = \frac{e^{-\frac{x-\mu}{s}}}{s(1+e^{-\frac{x-\mu}{s}})^2} = \frac{1}{4s}\text{sech}^2(\frac{x-\mu}{2s}) \tag{8}$$

The distribution average is calculated as the average of μ and variance $\frac{s^2\pi^2}{3}$. A unique feature of the logistic distribution is its bell-shaped curve similar to a normal distribution, but with the possibility of a kurtosis through scale variables (s). Figure 2 illustrates the logistic distribution with an average of 14.78% and scale of 0.22% selected in the distribution based on the actual measurement data for the capacity factor.

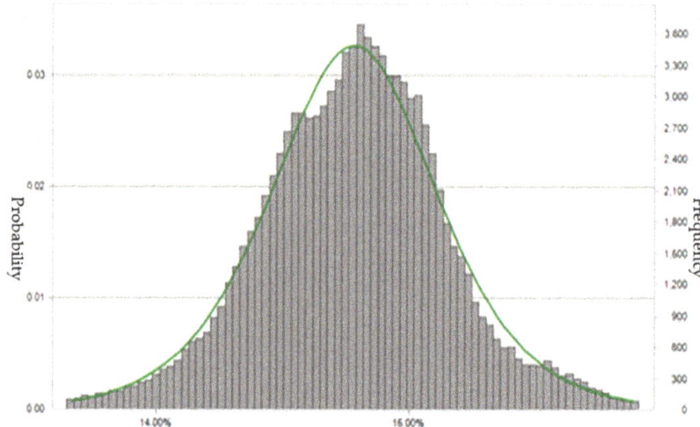

Figure 2. Probability distribution for capacity factor (logistic distribution).

4.1.2. Discount Rate

The triangular distribution was used in considering the uncertainty surrounding the characteristics of the discount rates. This distribution can be used when the MLE is accurate, and information on the maximum and minimum is certain. The PDF of the triangular distribution is provided as follows [32]:

$$f(x) = \frac{2(x-a)}{(m-a)(b-a)} \tag{9}$$

In the above formula, a refers to the minimum value; m, to the mode; and b, to the maximum value. The average (μ) and variance (σ^2) of this distribution are as follows:

$$\mu = \frac{a+m+b}{3} \tag{10}$$

$$\sigma^2 = \frac{(a^2+b^2+m^2-ab-am-bm)}{18} \tag{11}$$

According to the study conducted by Choi and Park [33], the discount rate reached a maximum of 7.5% in 2001 and currently equates to 5.5%, but requires a reduction of 1% in the future. Therefore, this study applied the triangular distribution of a minimum of 4.5%, a mode of 5.5%, and a maximum of 7.5%. Figure 3 shows the triangular distribution generated based on the above assumptions.

4.1.3. O&M Costs

The preceding study by the International Energy Agency (IEA) [34] was referenced in establishing the probability distribution for O&M costs, with the establishment of the normal distribution for this purpose. This distribution is applicable when the probability of MLE is high. The PDF of the normal distribution is presented as the following:

$$f(x) = \frac{1}{\sigma\sqrt{2\pi}}e^{-\frac{(x-\mu)^2}{2\sigma^2}} \tag{12}$$

In the above formula, μ refers to the average, while σ represents the standard deviation. In this study, the surveyed average for commercial facilities was KRW 37,365 (USD 33.97 (In this study, USD 1 = KRW 1100 is applied))/kW, and KRW 11,667 (USD 10.61)/kW for residential facilities. The standard deviation was assumed at an average of 5%, identical to the study by the IEA [34], with the standard deviations for commercial and residential facilities as KRW 1 868 (USD 1.70)/kW and KRW 583 (USD 0.53)/kW, respectively. Figures 4 and 5 are normal distributions based on the above-mentioned assumptions.

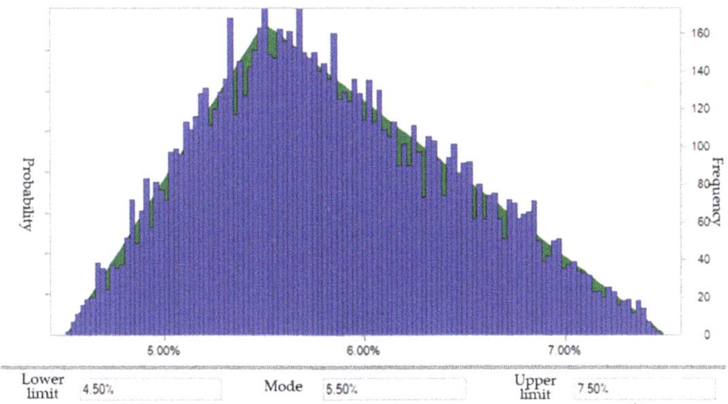

Figure 3. Probability distribution for discount rate (triangular distribution).

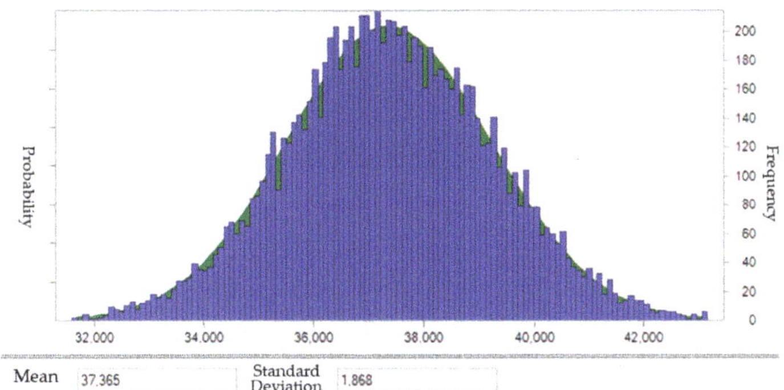

Figure 4. Probability distribution for commercial operations and maintenance (O&M) costs (normal distribution).

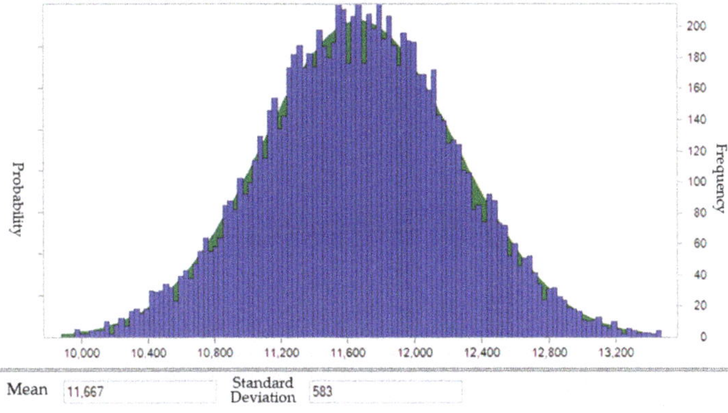

Figure 5. Probability distribution for residential O&M costs (normal distribution).

4.1.4. CAPEX

A normal distribution is assumed for the probability distribution for CAPEX, similar to the O&M costs as per above. One difference is the application of the standard deviation. The surveyed numbers are identical in that they were applied according to an average, but the standard deviation was assumed as 10% of the average. The reason for this is the greater uncertainty of CAPEX compared to O&M costs, caused by a reduction in costs resulting from technological advances or factors causing unexpected cost increases.

The CAPEX for commercial facilities recorded an average of KRW 1.6 million (USD 1 454.55)/kW, while the standard deviation totaled KRW 160,000 (USD 145.45)/kW. As for residential facilities, the average reached KRW 1.8 million (USD 1 636.36)/kW, while the standard deviation amounted to KRW 180,000 (USD 163.64)/kW. Figures 6 and 7 display the normal distributions for commercial and residential facilities.

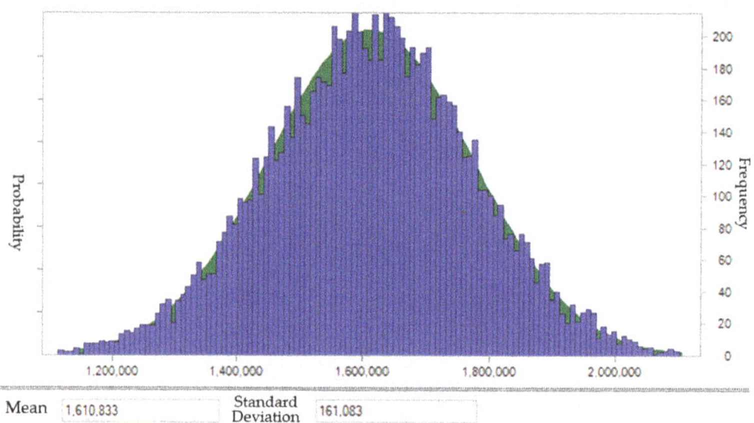

Figure 6. Probability distribution for commercial facility capital expenditure (CAPEX) (normal distribution).

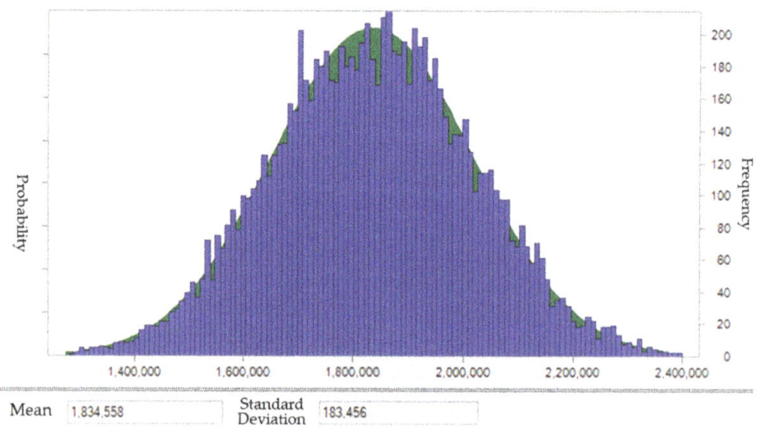

Figure 7. Probability distribution for residential facility CAPEX (normal distribution).

4.1.5. System Degradation Rate

The triangular distribution is assumed for the system degradation rate after referring to the preceding study by the IEA [34]. The technological threshold for photovoltaic energy generation is clear, meaning that the use of a triangular distribution seems reasonable. In IEA [34], a mode of 0.5% was applied, whereas this study applied 0.7%, as surveyed. The minimum and maximum values were applied at 0% and 0.8%, respectively, identical to the study by the IEA [34]. Figure 8 shows the triangular distribution generated based on the above assumptions.

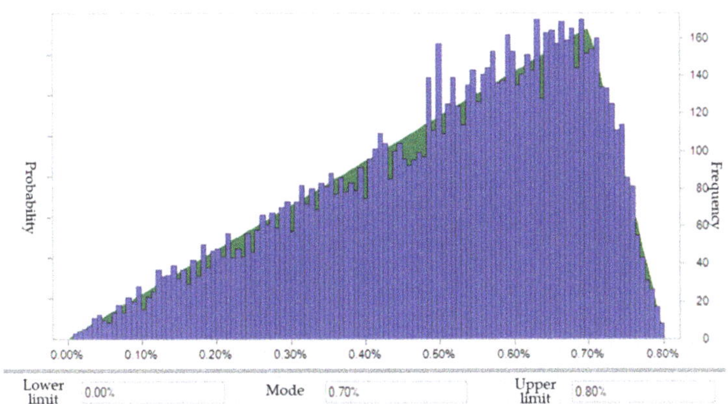

Figure 8. System degradation rate probability distribution (triangular distribution).

4.1.6. Corporate Tax

Corporate tax is a policy variable, and the following scenario was proposed as a hypothesis. The corporate tax will decline from its current state until it reaches a range where all corporate tax is exempt. The probability distribution suited to this hypothesis is the triangular distribution, which can arbitrarily establish a threshold.

This study assumes the triangular distribution, while the mode and maximum value were set at the current level of corporate tax at 24.2% (including residence tax), with the minimum value set at 0% (exemption of corporate tax scenario). Finally, as per Figure 9, a right triangle distribution was generated where the probability reaches its highest point at 24.2% and then declines gradually.

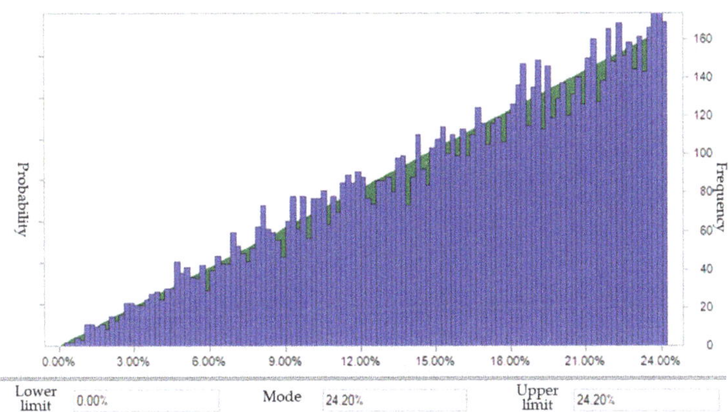

Figure 9. Corporate tax probability distribution (triangular distribution).

4.2. *Results of Stochastic Simulation*

This study uses the Monte Carlo simulation technique and encompasses the characteristics of variables that reflect the uncertainty and variability of solar PV generation in order to select a probability distribution. Arbitrary random numbers were generated within the distribution, and by repeating this process 10,000 times, this study estimates the range of meaningful solar LCOE with a confidence interval of 95%.

In the case of commercial solar energy, the LCOE records an average of KRW 159.49 (14 cents)/kWh with a standard deviation of KRW 13.31 (1 cent)/kWh. The 95% confidence interval was between KRW 133.60 (12 cents)/kWh and KRW 186.52 (17 cents)/kWh. Figure 10 shows the probability distribution for commercial solar energy generation, while Table 3 provides the relevant statistics.

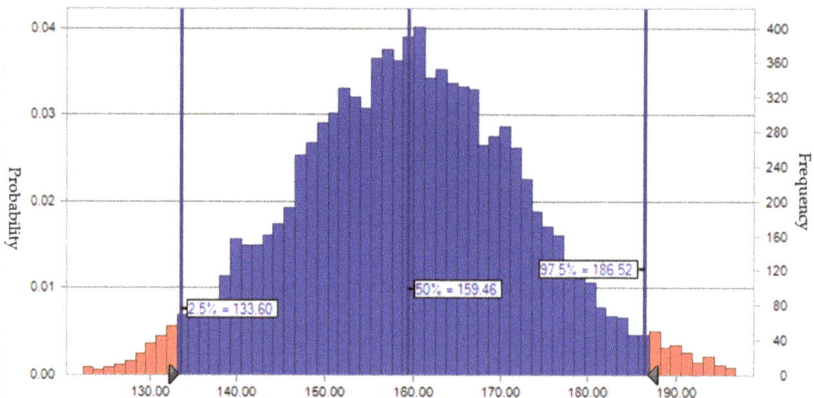

Figure 10. Probability distribution for commercial solar LCOE.

Table 3. LCOE statistics for commercial solar energy generation.

Statistics	Value	Statistics	Value
Reference value	165.97	Kurtosis	3.04
Average	159.49	Variation coefficient	0.0835
Median value	159.46	Minimum	114.84
Standard deviation	13.31	Maximum	216.08
Variance	177.29	Range width	101.24
Skewness	0.0647	Standard error	0.13

In the case of residential solar LCOE, an average of KRW 137.15 (12 cents)/kWh was recorded with a standard deviation of KRW 14.80 (1 cent)/kWh. The confidence interval at 5% significance level showed a minimum value of KRW 109.67 (10 cents)/kWh and a maximum of KRW 167.35 (15 cents)/kWh. Figure 11 and Table 4 illustrate the probability distribution and statistics for residential solar LCOE, respectively.

Table 4. Residential solar LCOE statistics.

Statistics	Value	Statistics	Value
Reference value	135.65	Kurtosis	2.97
Average	137.15	Variation coefficient	0.1079
Median value	136.75	Minimum	75.77
Standard deviation	14.80	Maximum	197.15
Variance	219.06	Range width	100.56
Skewness	0.1977	Standard error	0.15

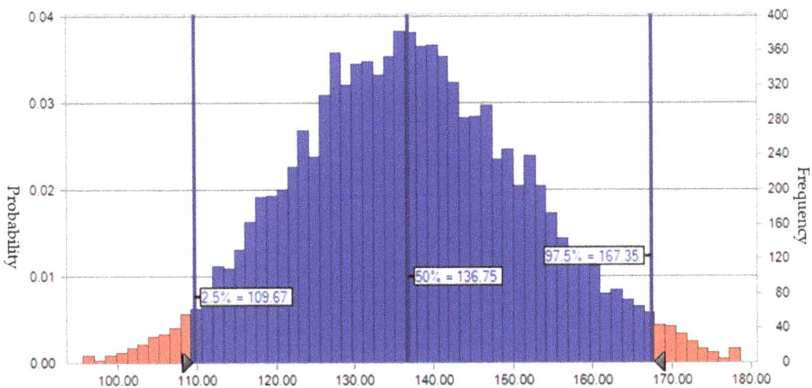

Figure 11. Probability distribution for the residential solar LCOE.

4.3. Sensitivity Analysis Results

Figures 12 and 13 show the results of the analysis on the variance contribution of the LCOE for solar PV. The factor with the greatest impact on the LCOE of commercial and residential solar PV is established as CAPEX. The contribution of this variable was recorded at 57.3% and 74.8% for commercial and residential facilities, respectively, while the proportion accounts for an absolute majority. This implies that CAPEX is the most common and significant factor to consider when seeking to improve the economics of solar power generation activities of both commercial and residential facilities. Therefore, it is essential to determine the factor that will reduce the relevant costs.

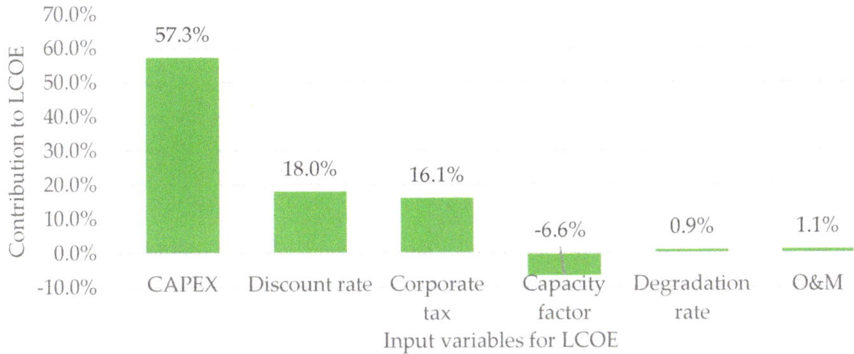

Figure 12. Sensitivity of the commercial solar LCOE.

The second most crucial factor to consider in improving the economy is the discount rate for commercial solar energy generation, which accounted for 18% of the variance contribution. Similarly, it was also the second-most important factor for residential generation facilities, accounting for 17.7% of the variance contribution. This implies that the development of a policy aimed at reducing the burden of financial costs may be effective.

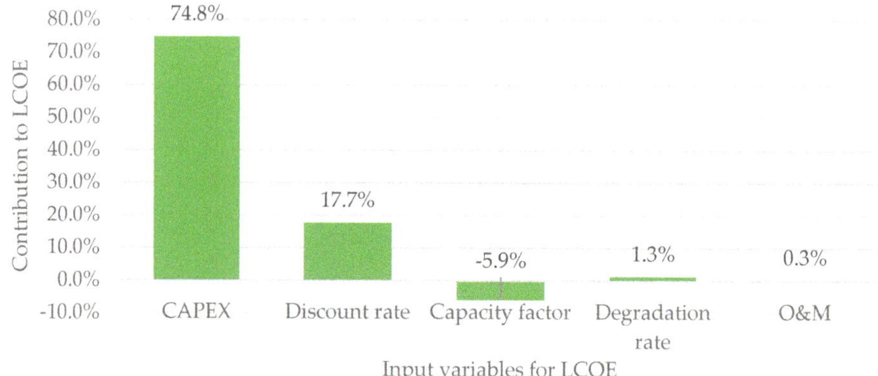

Figure 13. Sensitivity of the residential solar LCOE.

5. Discussion

The stochastic LCOE analysis methodology proposed in this study was applied to solar PV for residential and commercial use in South Korea. The deterministic LCOE performed in most existing studies does not indicate the uncertainty of the input variable. Although the sensitivity analysis of the deterministic methodology shows the uncertainty of input variables, the number of scenarios is limited. This study is meaningful in that the confidence interval of the results is derived by reflecting the stochastic characteristics of the input variable instead of finding the deterministic LCOE. This study also contributed to producing realistic analysis results by collecting the actual data generated in the solar PV project.

Among the input variables, the capacity factor was set to the optimal distribution function using the actual power generation data of solar PV. As a result of statistical analysis, significant statistical values, such as a reference value, an average, a median value, a standard deviation, a minimum value, and a maximum value, were derived. Commercial solar LCOE was estimated to range between KRW 115 (10 cents)/kWh and KRW 197.4 (18 cents)/kWh at a 95% confidence level. The median was valued at KRW 160.03 (15 cents)/kWh. The LCOE of residential solar was estimated to range between KRW 109.7 (10 cents)/kWh and KRW 194.1 (18 cents)/kWh at a 95% confidence level and a median value of KRW 160.03 (15 cents)/kWh. The sensitivity analysis showed that CAPEX had the most significant impact on solar LCOE, followed by the discount rate and corporate tax.

Therefore, to reduce the LCOE of solar PV, it is necessary to strive to reduce CAPEX. The increasing dissemination of solar PV could also lower CAPEX in PV. The reasons behind the high hardware and soft costs of solar PV facilities in Korea are twofold: the unique economic environment and lack of experience. For example, Germany is continuously lowering costs by steadily distributing solar PV systems across the country. If Korea followed suit and gained valuable learning experiences as a result, it is important to consider how much solar LCOE could be reduced. To answer this question, we assumed that Korea holds the equivalent level of knowledge and experience to Germany. Specifically, we assumed that Korea's hardware costs, including the costs of modules, inverters, connection bands, electric wiring, structures, and installation and construction, are reduced to the same level as they are in Germany. In addition, we assumed the same for soft costs, including supervisory costs, design costs, and general management costs, as well as the O&M costs (costs of replacing parts, safety management costs, etc.), excluding land leasing rates. Table 5 shows cost breakdown of a 100 kW solar system in Korea.

Table 5. Cost breakdown of a 100 kW solar system.

Items of Hhardware Costs	KRW	Items of Soft Costs	KRW	Items of O&M Costs		KRW
Modules	62,124,000	License and permits	9,000,000	Land lease costs		1,500,000
Inverters	14,375,000	Standard facility charges	8,390,000	Parts replacement costs	Inverters	718,750
Connection bands	2,200,000	Insurance premiums	1,141,623		Fuses, etc.	240,000
Electric wiring	601,678	Supervisory costs	1,500,000	Safety management costs		1,277,760
Structures	5,895,677	Other expenses	5,136,649	Total		3,736,510
Installation construction costs	23,933,435	Design costs	1,500,000			
Total	109,129,790	General management costs	6,924,483			
		Profits	5,570,428			
		Total	39,163,183			

Figure 14 indicates how Korea's solar LCOE could be reduced through the reduction of domestic installation costs for solar energy generation facilities to the same level as Germany. If done effectively, Korea's solar LCOE would drop KRW 26.6 (2 cents)/kWh. Similarly, if the soft costs and O&M costs are also reduced to Germany's level, the solar LCOE would be reduced by KRW 17 (1.5 cents)/kWh and KRW 6.3 (0.6 cents)/kWh, respectively. Therefore, if Korea's overall installation costs are reduced as per Germany's levels, the solar LCOE would decrease to KRW 97.2 (9 cents)/kWh. This is even less than Germany's current solar LCOE of KRW 122 (11 cents)/kWh, due to Korea's advantage in terms of capacity factor and corporate tax. However, decreasing installation costs cannot be achieved within a limited period of time. Germany managed to reduce costs based on experience gained through years of installing solar energy generation facilities. Therefore, if Korea continues to deploy solar energy systems, reduced solar LCOE below KRW 100 (9 cents)/kWh can be achieved.

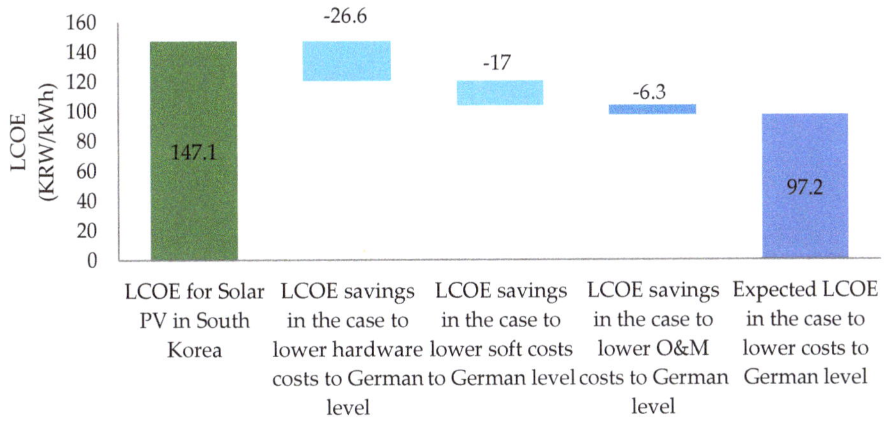

Figure 14. Commercial solar LCOE adjustment in South Korea after lowering costs to the level of Germany.

The higher cost of domestic modules and inverters in Korea also indicated that increased efforts are required to lower costs in the manufacturing sector. In addition, the high installation and construction costs are attributable to the country's high labor costs and extended construction periods. With limited scope to lower labor costs, efforts should be focused on reducing construction periods. Due to civil complaints often prolonging construction projects, a policy to increase public acceptance of solar energy generation facilities is also required.

Subsequently, we examine license and permit costs. License and permit costs paid to local autonomies in Korea are 10 times higher than in Germany and 50 times higher than in China. The greatest challenge for solar energy generators in Korea is securing licenses and permits for development activities from local autonomies. This urgently calls for the development of a policy to lower license and permit costs and to streamline related procedures. In addition, domestic grid connection costs in Korea are more than four times higher than in China, highlighting the need for lowering these costs. Domestic general management costs are also higher, exceeding those of other countries by more than 10 times. Cost reductions must be achieved by systemizing domestic projects.

This study also recommends the reduction of value-added tax on the installation of solar energy generation facilities. Recently, the National Energy Administration of China announced a policy to reduce taxes for solar generators, introducing a refund of 50% on value-added taxes for supplies associated with solar energy and the lowering of taxes on the use of farmlands. This policy will be adopted by 2020 to promote solar PV in the country. The reduction of corporate tax on the profits of solar energy generators should also be considered. The United States is accelerating the deployment of renewable energy sources through an investment tax credit and production tax credit, serving as adequate reference points.

Policy efforts to lower discount rates are also required. Discount rates comprise the cost of debt and the cost of equity, which can be lowered strategically. For instance, the cost of debt can be lowered by offering preferential interest rates for loans to generators of renewable energy. The promotion and support of renewable energy project financing (PF) should be additionally considered. The cost of debt can also be lowered through the government's active promotion of PF and support for the development of new investment products, with the objective to establish an industry ecosystem for renewable energy.

The promotion of renewable energy is an absolute necessity, considering the need to reduce GHG emissions and improve air quality. However, if the promotion of renewable energy becomes a burden to consumers in the form of excessive electricity bills, the deployment of renewable energy could be restricted. Should the cost of solar PV be reduced based on the results of this study, it would lead to a decrease in the unit cost of solar energy generation. Consequently, a reduction in power generation-related payments and, ultimately, relief in terms of electricity bills for consumers will follow.

The methodology proposed in this study is expected to be applicable to solar and other energy sources in other countries. A stochastic approach could be generalized if the methodology of this study is widely used in future studies. In addition, meaningful implications can be drawn from comparing countries and energy sources.

Author Contributions: C.-Y.L. designed the study, reviewed the related literature, and interpreted the results. J.A. outlined the methodology, and developed the model. All authors provided substantial writing contributions and significant comments on numerous drafts. All authors have read and agreed to the published version of the manuscript.

Funding: This work was supported by the Korea Energy Economics Institute (KEEI) grant funded by the South Korean Prime Minister's Office.

Conflicts of Interest: The authors declare no conflict of interest.

References

1. Olhoff, A.; Christensen, J.M. *The Emissions Gap Report 2017: A UNEP Synthesis Report*; The United Nations Environment Programme (UNEP) Press: Nairobi, Kenya, 2017.
2. Hwang, S.-H.; Kim, M.-K.; Ryu, H.-S. Real Levelized Cost of Energy with Indirect Costs and Market Value of Variable Renewables: A Study of the Korean Power Market. *Energies* **2019**, *12*, 2459. [CrossRef]

3. Poulsen, T.; Hasager, C.B.; Jensen, C.M. The role of logistics in practical levelized cost of energy reduction implementation and government sponsored cost reduction studies: Day and night in offshore wind operations and maintenance logistics. *Energies* **2017**, *10*, 464. [CrossRef]
4. Mendicino, L.; Menniti, D.; Pinnarelli, A.; Sorrentino, N. Corporate power purchase agreement: Formulation of the related levelized cost of energy and its application to a real life case study. *Appl. Energy* **2019**, *253*, 113577. [CrossRef]
5. Wang, X.; Kurdgelashvili, L.; Byrne, J.; Barnett, A. The value of module efficiency in lowering the levelized cost of energy of photovoltaic systems. *Renew. Sust. Energ. Rev.* **2011**, *15*, 4248–4254. [CrossRef]
6. Kasperowicz, R.; Pinczyński, M.; Khabdullin, A. Modeling the power of renewable energy sources in the context of classical electricity system transformation. *J. Int. Stud.* **2017**, *10*, 264–272. [CrossRef]
7. Bhandari, R.; Stadler, I. Grid parity analysis of solar photovoltaic systems in Germany using experience curves. *Sol. Energy* **2009**, *83*, 1634–1644. [CrossRef]
8. Branker, K.; Pathak, M.; Pearce, J.M. A review of solar photovoltaic levelized cost of electricity. *Renew. Sust. Energ. Rev.* **2011**, *15*, 4470–4482. [CrossRef]
9. Rhodes, J.D.; King, C.; Gulen, G.; Olmstead, S.M.; Dyer, J.S.; Hebner, R.E.; Beach, F.C.; Edgar, T.F.; Webber, M.E. A geographically resolved method to estimate levelized power plant costs with environmental externalities. *Energy Policy* **2017**, *102*, 491–499. [CrossRef]
10. Clauser, C.; Ewert, M. The renewables cost challenge: Levelized cost of geothermal electric energy compared to other sources of primary energy–Review and case study. *Renew. Sust. Energ. Rev.* **2018**, *82*, 3683–3693. [CrossRef]
11. Chadee, X.T.; Clarke, R.M. Wind resources and the levelized cost of wind generated electricity in the Caribbean islands of Trinidad and Tobago. *Renew. Sust. Energ. Rev.* **2018**, *81*, 2526–2540. [CrossRef]
12. Mundada, A.S.; Shah, K.K.; Pearce, J.M. Levelized cost of electricity for solar photovoltaic, battery and cogen hybrid systems. *Renew. Sust. Energ. Rev.* **2016**, *57*, 692–703. [CrossRef]
13. Nissen, U.; Harfst, N. Shortcomings of the traditional "levelized cost of energy" [LCOE] for the determination of grid parity. *Energy* **2019**, *171*, 1009–1016. [CrossRef]
14. Kore Energy Economics Institute. *An International Comparative Analysis on Levelized Costs based on Solar Cost Analysis*; KEEI Press: Ulsan, Korea, 2017. (In Korean)
15. Ragnarsson, B.F.; Oddsson, G.V.; Unnthorsson, R.; Hrafnkelsson, B. Levelized cost of energy analysis of a wind power generation system at Burfell in Iceland. *Energies* **2015**, *8*, 9464–9485. [CrossRef]
16. Castro-Santos, L.; Garcia, G.P.; Estanqueiro, A.; Justino, P.A. The Levelized Cost of Energy (LCOE) of wave energy using GIS based analysis: The case study of Portugal. *Int. J. Elec. Power* **2015**, *65*, 21–25. [CrossRef]
17. Lai, C.S.; McCulloch, M.D. Levelized cost of electricity for solar photovoltaic and electrical energy storage. *Appl. Energy* **2017**, *190*, 191–203. [CrossRef]
18. Kirchem, D.; Lynch, M.A.; Bertsch, V.; Casey, E. Modelling demand response with process models and energy systems models: Potential applications for wastewater treatment within the energy-water nexus. *Appl. Energy* **2020**, *260*, 114321. [CrossRef]
19. Charnes, J. *Financial Modeling with Crystal Ball and Excel + Website*; John Wiley & Sons: Hoboken, NJ, USA, 2012; Volume 757.
20. Månberger, A.A.; Stenqvist, B.A. Global metal flows in the renewable energy transition: Exploring the effects of substitutes, technological mix and development. *Energy Policy* **2018**, *119*, 226–241. [CrossRef]
21. Abd Alla, S.; Bianco, V.; Tagliafico, L.A.; Scarpa, F. Life-cycle approach to the estimation of energy efficiency measures in the buildings sector. *Appl. Energy* **2020**, *264*, 114745. [CrossRef]
22. Shi, Y.; Zeng, Y.; Engo, J.; Han, B.; Li, Y.; Muehleisen, R.T. Leveraging inter-firm influence in the diffusion of energy efficiency technologies: An agent-based model. *Appl. Energy* **2020**, *263*, 114641. [CrossRef]
23. Tvaronavičienė, M.; Prakapienė, D.; Garškaitė-Milvydienė, K.; Prakapas, R.; Nawrot, Ł. Energy efficiency in the long run in the selected European countries. *Econ. Sociol.* **2018**, *11*, 245–254. [CrossRef]
24. Adenle, A.A. Assessment of solar energy technologies in Africa-opportunities and challenges in meeting the 2030 agenda and sustainable development goals. *Energy Policy* **2020**, *137*, 111180. [CrossRef]
25. Klaniecki, K.; Duse, I.A.; Lutz, L.M.; Leventon, J.; Abson, D.J. Applying the energy cultures framework to understand energy systems in the context of rural sustainability transformation. *Energy Policy* **2020**, *137*, 111092. [CrossRef]

26. Ahmed, A.; Gasparatos, A. Multi-dimensional energy poverty patterns around industrial crop projects in Ghana: Enhancing the energy poverty alleviation potential of rural development strategies. *Energy Policy* **2020**, *137*, 111123. [CrossRef]
27. Lopez, A.R.; Krumm, A.; Schattenhofer, L.; Burandt, T.; Montoya, F.C.; Oberlaender, N.; Oei, P.-Y. Solar PV generation in Colombia—A qualitative and quantitative approach to analyze the potential of solar energy market. *Renew. Energy* **2020**, *148*, 1266–1279. [CrossRef]
28. Winston, W.L. *Simulation Modeling Using@ RISK*; Wadsworth Publising Company: California, CA, USA, 1996.
29. Barreto, H.; Howland, F. *Introductory Econometrics: Using Monte Carlo Simulation with Microsoft Excel*; Cambridge University Press: Cambridge, UK, 2006.
30. Spooner, J.E. Probabilistic estimating. *J. Constr. Div.* **1974**, *100*.
31. Alotaibi, R.M.; Rezk, H.R.; Ghosh, I.; Dey, S. Bivariate exponentiated half logistic distribution: Properties and application. *Commun. Stat. Theory Methods* **2020**, 1–23, in press. [CrossRef]
32. Stein, W.E.; Keblis, M.F. A new method to simulate the triangular distribution. *Math. Comput. Model.* **2009**, *49*, 1143–1147. [CrossRef]
33. Choi, J.; Park, D. An Estimation of the Social Discount Rate for the Economic Feasibility Analysis of Public Investment Projects. *J. Soc. Sci.* **2017**, *43*. (In Korean)
34. Richter, M.; Tjengdrawira, C.; Vedde, J.; Green, M.; Frearson, L.; Herteleer, B.; Jahn, U.; Herz, M.; Kontges, M.; Stridh, B. *Technical Assumptions Used in PV Financial Models*; International Energy Agency (IEA) Report; IEA Press: Paris, France, 2017.

Article

Quantifying Public Preferences for Community-Based Renewable Energy Projects in South Korea

Rahel Renata Tanujaya [1], Chul-Yong Lee [2], JongRoul Woo [3], Sung-Yoon Huh [4,*] and Min-Kyu Lee [5,*]

[1] Interdisciplinary Program of Science and Technology Policy, Pukyong National University, 365 Sinseon-ro, Nam-gu, Busan 48547, Korea; rahelrenata@pukyong.ac.kr
[2] School of Business, Pusan National University, 2 Busandaehak-ro 63beon-gil, Geumjeong-gu, Busan 46241, Korea; cylee7@pusan.ac.kr
[3] Graduate School of Energy and Environment (KU-KIST Green School), Korea University, 145 Anam-ro, Seongbuk-gu, Seoul 02841, Korea; jrwoo@korea.ac.kr
[4] Department of Energy Policy, Graduate School of Energy and Environment, Seoul National University of Science and Technology, 232 Gongneung-ro, Nowon-gu, Seoul 01811, Korea
[5] Graduate School of Management of Technology, Pukyong National University, 365 Sinseon-ro, Nam-gu, Busan 48547, Korea
* Correspondence: sunghuh@seoultech.ac.kr (S.-Y.H.); minkyu@pknu.ac.kr (M.-K.L.); Tel.: +82-2-970-6597 (S.-Y.H.); +82-51-629-5649 (M.-K.L.)

Received: 7 April 2020; Accepted: 8 May 2020; Published: 10 May 2020

Abstract: Under the new climate regime, renewable energy (RE) has received particular attention for mitigating the discharge of greenhouse gas. According to the third energy master plan in South Korea, by 2040, 30–35% of the energy demand must met with RE sources. To ensure relevant policy design to achieve this goal, it is crucial to analyze the public's willingness to accept community-based RE projects. This study conducted a nationwide survey to understand the opinion of the public and also that of local inhabitants living near a RE project. A choice experiment was employed to measure public preferences toward RE projects. The analysis reveals that the type of energy source, distance to a residential area, and annual percentage incentives could affect acceptance levels. Additionally, investment levels were a factor in local inhabitants' acceptance of energy-related projects. This study presents the relevant policy implications in accordance with the analysis results.

Keywords: choice experiment; renewable energy; willingness to accept; multinomial logit models

1. Introduction

Recently, many nations started orienting their energy policy initiatives to "the new climate regime" for mitigating the discharge of greenhouse gases. The United Nations asked many nations around the world to replace their traditional energy source with a more sustainable alternative. Each nation was required to hand in nationally decided contributions and formulate decrease goals quinquennially [1]. In this context, the South Korean government started to follow the global trend by replacing conventional energy sources such as coal with a renewable energy (RE) gradually.

The government has set an ambitious target that by 2040, 30–35% of the energy demand should be met using RE sources [2]. To meet this target, the level of social acceptance regarding RE needs to improve. Social acceptance of RE can be classified into the public acceptance and the local acceptance of inhabitants living near an RE farm. Acceptance is associated with the usual recognition caught by the greater part of people; it presents a challenge given that it is difficult to address in the short term [3]. For example, in South Korea, 37.5% of photovoltaic (PV) power and wind farms admitted in 2016 were canceled or postponed because of opposed local inhabitants [4]. One of the significant causes for local inhabitants' opposition to constructing RE projects is that RE projects provide no benefit to residents [5]. To address this issue, the South Korean government needs comprehensive information on public preferences for community-based RE projects. Such data allows the government to customize future RE incentives and policies in line with people's preferences.

The current study intends to determine the preferences toward RE projects in South Korea employing a choice experiment (CE) method. We consider five major attributes of RE projects: RE types, distance to residential areas, investment types, investment level, and annual percentage incentives. The marginal willingness to accept (MWTA) estimates can provide policymakers with the necessary information to stimulate social acceptance. This study is predominantly concerned with the marginal effect of the economic influence regarding community-based RE projects, covering three RE sources: wind, PV, and biomass. The economic impact is in the form of incentives that are needed for the public to accept RE plant projects. Finding the public's MWTA will be one of the goals of this study, differentiating it from other studies, which mainly focused on the public's marginal willingness to pay (MWTP) [6–9]. Several studies, such as that by Woo et al. [4] and Botelho et al. [10], also estimated the willingness to accept (WTA) for wind, solar, biomass, and hydropower power plants, but they used the contingent valuation method (CVM) for analysis.

The main questions addressed by this study are: what are factors that will affect the public's acceptance of RE projects and what are the financial implications in terms of the final product's value? Thus, both the general public's and local residents' MWTA for RE projects needs to be determined. The results will be useful for creating more appropriate community-based RE projects in South Korea. To do this, first, data on the standard of acceptance of both the public and local inhabitants is gathered and analyzed using a CE approach to determine the relationship between each observed alternative and the value of the goods [11]. Unlike the commonly used CVM, the CE approach is able to analyze the relationship between each factor and alternative in a one-step process. In choosing the best option for an RE source, respondents typically consider multiple aspects, causing a significant trade-off between choices. As a result, policymakers will be able to customize different aspects of their policy to favor everyone involved.

The rest of this study is structured as follows. Section 2 reviews the literature regarding the evaluation of public preferences for RE projects. In Section 3, we explain the CE approach to evaluate public preferences and the survey data. Section 4 depicts the empirical results, followed by a relevant discussion. Conclusions are made in the last section.

2. Literature Review

Public acceptance of RE projects can differ according to several determinants such as environmental factors, the location of the projects, and other socio-economic factors. For example, supporters and opponents of wind energy turbines tend to base their claims on the local and global environmental and economic factors [3]. Understanding of effects of wind turbines on health, environment, and community wellbeing will lead to RE project siting decision. The "Not In My Back Yard" (NIMBY) syndrome has been affecting different opinions and public acceptance rates of RE projects [12,13]. For RE projects, the most significant factor for acceptance is associated with the perceived qualities of the location [14]. Some studies on the improvement of the South Korean policy to fit the RE initiative provided detailed information regarding how to implement the government's goal in real life using the available RE technology [15,16]. However, those studies do not include public opinion and focus more on the technological aspect of the policy.

The Korean government has been concerned with the increase in the production of RE without considering the impacts or problems that might surface because of the lack of coordination and planning. Until now, the RE policy in South Korea has been a top-down scheme, and is effective as a temporary solution [17]; however, in the long run, bottom-up schemes with public/private involvement are needed [18–20]. Lack of coordination between the government, public, and private sectors has caused confusion during the investment and participation process. To ensure consistent programs that can be accepted by the public, collaboration between all stakeholders is important. Camarinha-Matos et al. [20] mentioned a smart energy grid in which customers are treated as partners, with shared obligations and benefits. Thus, each local community can be self-sufficient by generating its own energy. There have been many studies on energy self-sufficient communities or villages, especially in the United Kingdom (UK), as it was considered the solution to RE project local opposition [21–23].

As mentioned before, public participation is important to decrease any possibility of retaliation. Usually, there can be a gap between how the public and local inhabitants perceive an RE project. The general public's opinion usually depends only on the outcome of RE projects, without weighing the process through which that outcome was achieved. Local residents, by contrast, directly witness the process and face the consequences of their choices, leading them to often oppose RE projects [24,25]. To ensure local acceptance, Germany and the UK tried to implement a locally owned small-scale RE project policy that would benefit local residents [17,26,27].

Specifically, Salm et al. [28] in Germany and Masini and Menichetti [29] for the European Union, conducted a survey on the willingness to accept RE projects. These studies indicated the period of investment return plays an important role in respondents' willingness to accept. There are several RE project case studies in Finland [30], the UK [31], Greece [25], and Spain [7]. Kosenius and Ollikainen [30] estimated the willingness to pay (WTP) for RE sources and considered effects of energy production as biodiversity, jobs, CO_2 emissions, and electricity bills. Scarpa and Willis [31] investigated public WTP for RE projects in UK using a CE approach. Dimitropoulos and Kontoleon [25] conducted a CE method to estimate preferences regarding wind farm projects in Greece. Moreover, Álvarez-Farizo and Hanley [7] tried to measure preferences for the environmental effects of wind plants in Spain using a CE method. According to their findings, social costs of environmental effects will relate to wind plant projects.

Most previous community-based RE project studies are from Europe and currently, the amount of research in Asia regarding this topic is little to non-existent. To the best of the authors' awareness, there have been few applications of a CE method to evaluate the willingness of local residents to accept for community-based RE projects in South Korea. Unlike other studies, this study attempts to compare public and local residents' preference for RE projects. Moreover, it is difficult to discover previous studies for quantitative analysis regarding policy impacts of financial incentives for RE projects. Therefore, this paper will give insightful information regarding community-based RE projects in Asian countries.

3. Methodology and Data

3.1. CE Survey

We gathered data regarding South Korean people's stated preferences for new RE power plant projects (solar PV, wind, and biomass power plants) employing a CE survey. The reason for using a CE survey is mainly to understand the respondent's attitude that causes them to favor specific alternatives [32]; the analysis result can inform policy design. This approach can identify the economic value of each attribute that makes up the alternatives and the total value of the alternatives [11]. This approach can provide insights for policymakers to customize their policies and ensure they are more suited to public preferences [33,34].

To design an appropriate CE for South Korean people's preferences for new RE power plant projects, we need to check the major attributes of RE power plant projects and assign relevant levels. It demands respondents to select the preferable alternative out of an assumed set of projects made from major attributes stated at some standards [35]. Table 1 depicts the attributes which can affect South Korean people's preferences for RE projects. It is assumed that other latent attributes are the same through all alternatives.

Specifically, we assumed that five attributes could describe RE projects. First, we included three technology options such as solar PV, wind, and biomass power plants. The reason for choosing these three levels is to analyze differences in preferences by various RE technologies. As of the time of the survey, in 2017 in South Korea, the RE sources for power generation are solar photovoltaic (7056 GWh/year produced in 2017), wind power (2169 GWh), hydro power (2819 GWh), ocean (489 GWh), bio (7467 GWh), and waste (23,867 GWh) [36]. Therefore, the three power sources included in the CE survey account for a substantial share of renewable power generation in South Korea. Moreover, the government designs to decrease the portion of waste energy in the long term [37], and hydro power is not very suitable to be promoted as a community-based project due to its big size and long gestation period. Ocean energy cannot be developed for private use, for which the application of community-based project is not suitable.

Second, the distance between the respondent's residential area and the renewable power plant was considered to be one of the key characteristics of the RE project. We assumed that the possible distance ranges from 100 to 1000 m. Those levels reflect the regulations on separation distance (i.e., setback distance) of RE power plants in several countries including South Korea. As of June 2018, 94 of the 228 local governments in South Korea enforced the separation distance regulation on photovoltaic power facilities, and most of them span 100 to 1000 m [38]. In terms of wind power, there are no statutory regulations in South Korea yet, but in general, it is suggested that a distance of 500 to 1500 m from residential area is negotiable with local residents [39]. Regulations and recommendation regarding the separation distance between the wind turbine and residential areas in other countries such as Germany, Canada, and the UK also extends around 300 and 1000 m [40].

Third, how respondents participate (invest) in the project determines the characteristics of the RE projects. Bond investment is a form of investment in an energy project, and a certain amount of interest is guaranteed each month, based on the amount of the investment. On the other hand, equity investment is when the investor share in both the profit and risk by owning a portion of the RE project as a shareholder. These two levels are representative methods of resident participation that are actually used in the community-based RE project in South Korea [4]. In addition, when promoting a community-based RE project, the South Korean government is planning to diversify the residents' participation method from existing equity participation to bonds, funds, etc. [41].

Fourth, the levels of respondents' participation can make a difference among RE projects. Recent literature highlighted that high level of public participation and engagement is a key factor in raising the acceptance toward local RE projects [42–45]. In particular, as the main purpose of a community-based RE project is to increase local acceptance by securing high levels of public participation and the level of engagement of residents can be different [21,46], it is important to examine people's preferences for this attribute. We assumed that there are two levels of respondents' participation. When the level of participation of respondents is low, respondents will only participate in procedures such as consent for construction and operation according to administrative requirements. However, when the level of respondents' participation is high, respondents participate directly in the operation and management of the RE project.

Lastly, the expected rate of return can vary according to the forms of RE projects. This attribute serves as a payment vehicle, and corresponds to the price attribute of the typical CE questionnaire. It is also the basis for WTA calculation. In an ongoing community-based RE project, local residents would receive a share of the benefit. Therefore, how much profit is shared is a crucial determinant of the acceptance of the RE project. Previous studies also confirmed that the financial incentives such as benefit sharing influenced people's attitude toward RE projects [47,48]. In addition, when using the stated preference techniques, it is significant to use a payment vehicle that is similar to the real world decision, which is the case for this attribute [49]. Based on previous RE projects in South Korea, the expected rate of return for each participant was assumed to range between 2%/year and 6%/year [50,51]. It was also considered that the rate of return on 3-year to 10-year government bonds ranged from 1.8% to 2.28% in 2017.

Table 1. Attributes of Renewable Energy (RE) projects for the choice experiment.

Attribute	Attribute Level
RE technology	(1) Solar photovoltaic (2) Wind (3) Biomass
Distance from residence	(1) 100 m (2) 500 m (3) 1000 m
Participation form	(1) Bond investment (2) Equity investment
Participation level	(1) Low (2) High
Expected rate of return	(1) 2%/year (2) 4%/year (3) 6%/year

We derived the 108 possible combinations from attributes and their levels in Table 1. Through Statistical Package for the Social Sciences (SPSS) 19.0.1, however, we selected only 18 orthogonal alternatives adopting a fractional factorial design to ensure the orthogonality of each attribute. Such alternatives are separated into six choice sets consisting of three alternatives and one status quo option. To save time needed for the respondents to respond, the respondents were classified into two groups, with each group demanding to answer three different sets. Then, the respondents are permitted to select the most preferred alternative from three alternatives and one status quo option in each choice set (see Table 2).

Table 2. A sample choice set.

	A	B	C	D
RE technology	Wind	Solar Photovoltaic	Biomass	No interest to participate (Status quo)
Distance from residence	500 m	1000 m (1 km)	100 m	
Participation form	Equity investment	Equity investment	Bond investment	
Participation level	Low	Low	High	
Expected rate of return	4%/year	6%/year	2%/year	
Most preferable option		V		

This study attempts to compare public acceptance with the local acceptance of the inhabitants currently living near the RE power plant. Thus, we conducted two separate CE surveys: (1) one for the South Korean nationwide public and (2) another for local inhabitants who reside within 1 km proximity of a RE power plant. The features of the survey design are represented in Table 3. Table 4 condenses the features of the 508 respondents in the public survey and 306 respondents in the local resident survey. The considerable difference in education level between the two samples needs to be addressed. It is probably due to the difference between the sampling method and survey method for the two samples, as shown in Table 4. This is a limitation of this survey and needs to be corrected in subsequent studies.

Table 3. Summary of survey design.

Type	(1) Survey for General Public	(2) Survey for Local Residents
Population	Head of household (and spouse), aged 20 to 65, nationwide	Head of household (and spouse), aged 20 to 65, living in administrative areas within 1 km of RE power plant grounds
Sample size	508 persons	306 persons
Sampling method	Sampled at random from proportional quotas based on age and region	Purposive quota sampling method
Method	Web survey	Face-to-face interview
Period	May 22 to May 29, 2017	May 19 to May 30, 2017
Survey firm	Hankook Research	

Table 4. Descriptive statistics of the respondents.

Type	Definition	(1) Survey for General Public	(2) Survey for Local Residents
		No. Respondents (%)	
Total		508 (100%)	306 (100%)
Gender	Male	244 (52%)	155 (50.7%)
	Female	264 (8.5%)	151 (49.3%)
Age	19–29	43 (8.5%)	12 (3.9%)
	30–39	103 (20.3%)	48 (15.7%)
	40–49	147 (28.9%)	80 (26.1%)
	50–59	151 (29.7%)	107 (35.0%)
	≥60	64 (12.6%)	59 (19.3%)
Education level	Less than high school	88 (17.3%)	220 (71.9%)
	More than college	420 (82.7%)	86 (28.1%)
Type of RE power plant	Wind power		101 (33.0%)
	PV		103 (33.7%)
	Biomass		102 (33.3%)

3.2. Model

We employed multinomial logit models to explore the South Korean public and local inhabitants' preferences for new RE power plant projects. The multinomial logit model is constructed based on random utility theory. The multinomial logit model is the most widely used discrete choice model. Its popularity is caused by the fact that the equation for the choice probabilities has a closed shape [32]. Although its IIA property is inappropriate in some choice situations, the power and applicability of logit models to represent choice behavior has been demonstrated in several studies in the field of energy [52,53]. The utility of an individual n by selecting alternative j in choice condition t U_{njt} is specified as Equation (1) [32,54].

$$U_{njt} = V_{njt} + \varepsilon_{njt} = \beta' x_{jt} + \varepsilon_{njt}, \tag{1}$$

The utility can be separated into the deterministic portion V_{njt} and the stochastic portion ε_{njt}. x_{jt} indicates the observable attributes for alternative j's attribute value in choice condition t, and β is the parameter for the equivalent attribute. Specifically, in this study, we defined the deterministic part of the utility function as Equation (2):

$$V = \beta_1 d_{wind} + \beta_2 d_{solar} + \beta_3 d_{biomass} + \beta_4 x_{distance} + \beta_5 d_{equity-invest} + \beta_6 d_{low-participant} + \beta_7 x_{return}, \tag{2}$$

where d_{wind}, d_{solar}, and $d_{biomass}$ are dummy variables representing wind, solar PV, and biomass energy technologies, respectively. Looking at the RE technology attribute in a sample choice set (Table 2), alternative A is wind power, B is solar photovoltaic, C is biomass, and alternative D is no-choice (opt-out) option. Our CE is designed so that this order was always fixed in all choice sets. For this reason, the alternative specific constant was not involved in the model. In addition, according to this, the base for estimation of the RE technology dummy variable were set to status quo, which means that dummy represents the difference between choosing alternative D. $x_{distance}$ represents the distance between the respondent's residential area and the renewable power plant. $d_{equity-invest}$ is a dummy variable representing equity investment, and $d_{low-participant}$ is a dummy variable representing a low level of respondent participation. Finally, x_{return} represents the expected rate of return.

Depending on the assumption that respondent selects an alternative from each choice situation to maximize their utility, the choice probability that respondent n selects alternative j in a choice condition t is defined as Equation (3).

$$P_{njt} = \Pr\left(U_{njt} > U_{nkt}, \forall k \neq j\right) = \Pr(\varepsilon_{nkt} - \varepsilon_{njt} < V_{njt} - V_{nkt}, \forall k \neq j), \tag{3}$$

If ε_{njt} follows an independent and identically distributed type I extreme value distribution, the choice probability that a respondent n will select alternative j in choice condition t can be expressed as Equation (4) [54].

$$P_{njt} = \frac{e^{V_{njt}}}{\sum_k e^{V_{nkt}}}, j = 1, \ldots, J, \tag{4}$$

The model can be solved easily by employing the maximum likelihood estimation, and the log-likelihood function for the estimation is specified as Equation (5).

$$LL = \sum_{n=1}^{N} \prod_{t=1}^{T} P_{njt}, \tag{5}$$

The MWTA can be calculated on the basis of the estimation results of β. The MWTP means how much a respondent is willing to accept for a unit change in the level of an attribute. The value can be computed by dividing each attribute's parameter estimate by the return attribute parameter estimate as Equation (4) [32].

$$MWTA_a = -\frac{\partial U/\partial x_a}{\partial U/\partial x_{return}} = -\frac{\hat{\beta}_a}{\hat{\beta}_6}, \tag{6}$$

$\hat{\beta}_a$ and x_a represent the mean estimated parameters and attribute values of the attributes, excluding the rate of return attribute. $\hat{\beta}_6$ and x_{return} indicate the estimated parameter and the attribute level for the rate of return attribute, respectively.

4. Results and Discussion

As noted before, this study divided the sample into two groups of respondents to analyze the preference gap between the public and the local residents for the community-based RE project. The multinomial logit models for both samples were analyzed using the NLOGIT 4.0 software, and the analysis results are shown in Table 5. For the estimation of the model without covariates, we used Equation (2) to analyze the data. As mentioned in Section 3.2, the MWTA for each attribute was calculated using Equation (4), and are also presented in Table 5. Discussion on the MWTA suggested below is only focused on the results that are statistically meaningful at either the 5% or 1% confidence level.

Table 5. Estimation results of multinomial logit models.

Variable		General Public		Local Residents	
		Coefficient (Standard Error)	MWTA (%)	Coefficient (Standard Error)	MWTA (%)
Renewable energy technology	Wind	−1.2211 ** (−0.1856)	10.3197	−2.9697 ** (0.2475)	15.3282
	Solar	0.4214 * (0.1695)	−3.5618	−1.8350 ** (0.2236)	9.4717
	Biomass	−0.9040 ** (0.1808)	7.6395	−2.8115 ** (0.2313)	14.5115
Distance from residence		0.0005 ** (0.9519)	−0.0042	0.0006 ** (0.0002)	−0.0031
Participation form (Equity)		−0.0680 (0.0964)	0.5750	−0.0694 (0.1274)	0.3580
Participation level (Low)		0.0454 (0.0808)	−0.3837	−0.2588 * (0.1180)	1.3360
Expected rate of return		0.1183 ** (0.0287)		0.1937 ** (0.0373)	
		N = 1524; Pseudo R^2 = 0.01 LL = −1750.080		N = 918; Pseudo R^2 = 0.022 LL = −970.149	

*: Statistical significance at 5% level; **: Statistical significance at 1% level; the pseudo R^2 merely has meaning when compared to another pseudo R-squared of the same type, on the same data, predicting the same outcome [55].

On the basis of the results of Table 5, more specifically, the sign of the estimated coefficients and the calculated MWTA, the differences in preferences between the two groups for the community-based RE project are discussed below. First, looking at the preference for RE technology attributes, all three coefficients of technology attributes are highly significant for both local residents and the general public with negative (−) signs, except for the solar PV power plant in the nationwide survey. The solar PV coefficient in the national survey shows a positive (+) sign with a 5% significance level. When estimating the dummy variable as the base level, the RE plant is not in the vicinity. The negative (−) sign basically means that people do not prefer renewable facilities around their residence. However, the general public shows a preference for solar PV power plants around their residence, even if there is no incentive, showing a favorable attitude toward solar PV technology. The relative preferences for the three RE technologies are solar PV, biomass, and wind power in both groups. In particular, the findings indicate that the preference gap between solar power and the other two RE sources was considerable. The differences in people's preferences (and willingness to pay) for different RE sources were already suggested in several studies, and this study reaffirms the existence of such differences. In particular, many existing studies found that people prefer solar PV technology over other RE alternatives [56–58], and this study supports it. In addition, in previous studies, the relative preference between wind power and biomass showed somewhat different results depending on the region and other factors (e.g., Kosenius and Ollikainen [30]). In this study, it was found that the South Korean people (both general public and local residents) prefer biomass to wind power, but this is not a big difference. The finding that solar PV is the preferred RE source of the South Korean people is consistent with previous studies [4,59].

Next, we discuss each group's MWTA for the RE technology attribute. First, we present the general public's MWTA. For the renewable technology type, the data indicates the public is willing to accept wind farm projects if they were given a 10% level of the annual expected rate of return and biomass plant if they were given 7.6%. However, for solar PV power plants, the general public appears to have an opinion that the project can be promoted even if the expected return is negative (−). However, it is impossible to establish a solar PV power plant at an investment loss for the public. Therefore, it can be assumed that the general public has a strong preference for solar PV technology.

Next, we will look at the local residents' MWTA estimations. The MWTA for all technology types is statistically significant at the 1% confidence level, showing local residents accept wind farms at a 15% expected rate of return, while solar PV and biomass plants show a lower annual rate of return (9.5% and 14.5%). When comparing the MWTAs between the two groups, the local residents had a higher MWTA for all three technologies. This means that a higher expected rate of return must be provided to ensure local residents' acceptance of RE projects, i.e., the local acceptance of community-based RE projects is lower than that of the general public. This finding confirm previous studies that identified various reasons for the low acceptance, such as NIMBY, social gap, and the difference in public versus private preferences [60,61]. Such a relatively lower acceptance of local residents compared to the general public is consistent with the results of Lee et al. [59], who analyzed the South Korean case. Therefore, projects need to consider these preferences when promoting community-based RE projects in South Korea. Thus, it will be necessary to focus more on planning and design from the perspective of local residents.

The interpretation of the distance attribute is as follows. For the distance, the general public's and local residents' opinions show a positive (+) sign and are meaningful at the 1% level. This indicates that it is preferred that the RE power plant is located further away from the respondents' residential area. The distance, which is also statistically meaningful at the 1% level, shows the public's MWTA of −0.0042%. From this result, we can assume that the public is willing to receive 0.0042% less annual expected rate of return for every 1 m the plant is located further away from their residential area. Additionally, local residents are willing to accept 0.0031% less for their annual rate of return if the plant is located 1 m further. In sum, even for a community-based RE project, the plant should be as far away from the densely populated area as possible, which leads to a positive response with a lower rate of returns.

Contrary to our expectation, the participation survey for the community-based RE project showed no significant impact on the preferences of both groups. Meanwhile, with regard to the expected rate of return attribute, we can examine that the results are statistically significant at the 1% level with the positive (+) sign. This coefficient, used in Equation (4), counts the respondents' MWTA. Considering that the sign of the expected rate of return is positive (+) and significant, it seems that the South Korean people are indifferent to investment methods such as stocks and bonds (which is related to the degree of investment risk and voting rights) while focusing mainly on return on investment.

Finally, with regard to the participation level attribute, there are various types of public participation in RE projects [42,43]. In this study, it is necessary to clarify that the participation level attribute refers to procedural participation. This is because financial participation is already reflected in the form of the expected rate of return attributes. First, in the case of the general public, it was found that the estimated coefficient of participation level attribute was not statistically significant. In contrast to the general public's result, the local residents' participation level has a negative coefficient and is statistically meaningful at the 5% confidence level with an MWTA of 1.3%. This means that the local residents will require a 1.3% higher annual rate of return for a low participation level. Recent studies reported the positive role of public participation on improving public acceptance of RE projects [43–45]. As community-based RE projects in South Korea are likely to involve local residents, it is necessary to increase the participation level of local residents to promote a successful community-based RE project. As the relatively high participation level showed to have a moderate impact on the MWTA (1.33%), significant financial savings can also be achieved.

In summary, the results indicate that the RE technology types, distance, and expected rate of return could affect the public acceptance of community-based RE projects, while the local residents' acceptance level could be affected by the similar attributes in addition to participation level. In addition, four main policy implications can be suggested from the analysis results. Firstly, in South Korea, the local acceptance of community-based RE projects is lower than that of the general public, so it is necessary to always keep in mind the acceptance gap between the two groups when working on new community-based RE projects in the future. Secondly, because the South Korean people prefer solar

PV above to the other two technologies and the preference gap is significant, it is recommended to disseminate community-based RE projects mainly focusing on solar PV power plants. With solar PV plants, it is expected that effective and smooth dissemination of community-based RE projects can be achieved with less financial input. The findings indicate the largest influence on the expected rate of return (WTA) was the selection of the renewable power source (for example, in the case of local acceptance, if the wind power or biomass technology is chosen rather than solar PV, an additional rate of return of approximately 5% to 6% should be provided to offset the potential resistance from local residents). Thirdly, the range between the plant and the residence remains an important attribute. In other words, although the community-based RE project is more resident-friendly than a conventional RE project, it is wise to deviate from the densely populated area as much as possible when selecting a site. Lastly, it was confirmed that procedural participation is as important as financial participation. It is recommended that similar studies, such as Langer et al. [42], Koch and Christ [62], and Liu et al. [63], should be conducted to determine the most effective local residents' participation plans for South Korea.

5. Conclusions

Greenhouse gas emissions rapidly increased since the first industrial revolution, resulting in a significant rise in global temperatures. The United Nations asked countries around the world to replace their conventional energy sources with a more sustainable option. Accordingly, the South Korean government has set a goal that by 2040, 30–35% of the domestic energy demand will be met with RE sources. To achieve this goal, social acceptance, including public acceptance and local acceptance as well as technological development, is crucial. Public opinion must be considered in order to develop RE policies. One way to understand public opinion and apply it to the policy frameworks is through a survey. The data collected in this survey can then be analyzed using several methods to understand the factors that affect public acceptance of RE. This study conducted a conjoint analysis CE to determine public acceptance of RE projects in South Korea.

We use the multinomial logit model to explore the survey data. The results show that the energy source type, distance, and annual percentage incentives significantly affect social acceptance. In addition, it was confirmed that public acceptance of RE was higher than local acceptance. These results are presumed to be due to the NIMBY syndrome. The result also shows that there is an investment level that only has an effect on local residents' acceptance but is not statistically significant in the general public's opinion. Local residents preferred solar power over both wind, which generates noise, and biomass power plants, which emit pollutants.

The government has the ability to create a better world by using policy instruments to convince the public and create a beneficial symbiosis to find solutions to both national and global challenges. These instruments can be in the form of economic support, which can be broadly divided into price discount (subsidies) or investment returns. In this study, we found that an appropriate investment return level is required to secure public acceptance and local acceptance. Local residents prefer a higher investment return than national residents do. Therefore, the government can create a policy program based on the distance between the project and the residential areas. As indicated in this study, the closer the distance, the greater is the investment return required.

However, there are other possible policy programs that can be created by customizing them to include every statistically significant attribute that would increase public and local acceptance. We could not sufficiently stress the importance of the public's participation and active promotional/educational programs to enhance the reputation of RE plants, which will help boost the acceptance rate of RE projects all over the country. Because countries differ in culture, mindset, and renewable technology levels, a careful approach is required to apply the results of this study to other countries. We expect that the general public and local resident's preference for RE will vary from country to country. Similar studies can be applied to other countries to conduct a comparative analysis.

This study hypothesized that the difference in acceptance for RE between general public and local residents is caused by living around RE plants. However, national and local residents may show different results depending on their gender, income and education levels. This study did not take into consideration these socio-economic attributes, leaving this issue for future study such as cross effects analysis with the demographic variables or latent class analysis. By including the socio-economic background of the respondents, it is possible to comprehend the general attitude of the respondents. The abovementioned recommendation will be beneficial to reduce public resistance.

Author Contributions: All the authors made contribution to this paper. R.R.T. suggested the significant ideas and estimated the survey data. C.-Y.L. designed the study and made the survey questionnaire. J.W. outlined the methodology and developed the model. S.-Y.H. explained the results. M.-K.L. investigated the research background and reviewed the related literature. All authors have read and agree to the published version of the manuscript.

Funding: This work was supported by the Pukyong National University Research Fund in 2017 (C-D-2017-1003).

Conflicts of Interest: The authors declare no conflict of interest.

References

1. Lee, C.Y.; Lee, M.K.; Yoo, S.H. Willingness to pay for replacing traditional energies with renewable energy in South Korea. *Energy* **2017**, *128*, 284–290. [CrossRef]
2. MOTIE. *The Third Energy Master Plan*; MOTIE: Sejong, Korea, 2019. (In Korean)
3. Petrova, M.A. From NIMBY to acceptance: Toward a novel framework-VESPA-For organizing and interpreting community concerns. *Renew. Energy* **2016**, *86*, 1280–1294. [CrossRef]
4. Woo, J.R.; Chung, S.; Lee, C.Y.; Huh, S.Y. Willingness to participate in community-based renewable energy projects: A contingent valuation study in South Korea. *Renew. Sustain. Energ. Rev.* **2019**, *112*, 643–652. [CrossRef]
5. Aitken, M. Wind power and community benefits: Challenges and opportunities. *Energy Policy* **2010**, *38*, 6066–6075. [CrossRef]
6. Kim, J.; Park, S.Y.; Lee, J. Do people really want renewable energy? Who wants renewable energy? Discrete choice model of reference-dependent preference in South Korea. *Energy Policy* **2018**, *120*, 761–770. [CrossRef]
7. Álvarez-Farizo, B.; Hanley, N. Using conjoint analysis to quantify public preferences over the environmental impacts of wind farms. An example from Spain. *Energy Policy* **2002**, *30*, 107–116. [CrossRef]
8. Vecchiato, D.; Tempesta, T. Public preferences for electricity contracts including renewable energy: A marketing analysis with choice experiments. *Energy* **2015**, *88*, 168–179. [CrossRef]
9. Sagebiel, J.; Müller, J.R.; Rommel, J. Are consumers willing to pay more for electricity from cooperatives? Results from an online Choice Experiment in Germany. *Energy Res. Soc. Sci.* **2014**, *2*, 90–101. [CrossRef]
10. Botelho, A.; Pinto, L.M.C.; Lourenço-Gomes, L.; Valente, M.; Sousa, S. Social sustainability of renewable energy sources in electricity production: An application of the contingent valuation method. *Sustain. Cities Soc.* **2016**, *26*, 429–437. [CrossRef]
11. Merino-Castello, A. Eliciting Consumers Preferences Using Stated Preference Discrete Choice Models: Contingent Ranking versus Choice Experiment. *UPF Econ. Bus. Work. Paper* **2003**, *705*. [CrossRef]
12. Bell, D.; Gray, Y.; Haggett, C. The 'social gap' in wind farm siting decisions: Explanations and policy responses. *Environ. Politics* **2005**, *14*, 460–477. [CrossRef]
13. Wüstenhagen, R.; Wolsink, M.; Bürer, M.J. Social acceptance of renewable energy innovation: An introduction to the concept. *Energy Policy* **2007**, *35*, 2683–2691. [CrossRef]
14. Wolsink, M. The research agenda on social acceptance of distributed generation in smart grids: Renewable as common pool resources. *Renew. Sustain. Energy Rev.* **2012**, *16*, 822–835. [CrossRef]
15. Kim, S.; Lee, H.; Kim, H.; Jang, D.H.; Kim, H.J.; Huh, J.; Cho, Y.S.; Huh, K. Improvement in policy and proactive interconnection procedure for renewable energy expansion in South Korea. *Renew. Sustain. Energy Rev.* **2018**, *98*, 150–162. [CrossRef]
16. Park, J.; Kim, B. An analysis of South Korea's energy transition policy with regards to offshore wind power development. *Renew. Sustain. Energy Rev.* **2019**, *109*, 71–84. [CrossRef]
17. Allen, J.; Sheate, W.R.; Diaz-Chavez, R. Community-based renewable energy in the Lake District National Park—local drivers, enablers, barriers, and solutions. *Local Environ.* **2012**, *17*, 261–280. [CrossRef]

18. Kolk, A.; van den Buuse, D. In search of viable business models for development: Sustainable energy in developing countries. *Corp. Gov.* **2012**, *12*, 551–567. [CrossRef]
19. Kellett, J. Community-based energy policy: A practical approach to carbon reduction. *J. Environ. Plan. Man.* **2012**, *50*, 381–396. [CrossRef]
20. Camarinha-Matos, L.M.; Afsarmanesh, H.; Boucher, X. The role of collaborative networks in sustainability. In Proceedings of the Working Conference on Virtual Enterprises, St. Etienne, France, 11–13 October 2010; pp. 1–16.
21. Rogers, J.C.; Simmons, E.A.; Convery, I.; Weatherall, A. Public perceptions of opportunities for community-based renewable energy projects. *Energy Policy* **2008**, *36*, 4217–4226. [CrossRef]
22. Khan, M.I.; Chhetri, A.B.; Islam, M.R. Community-based energy model: A novel approach to developing sustainable energy. *Energy Sources Part B* **2007**, *2*, 353–370. [CrossRef]
23. Ma, W.; Xue, X.; Liu, G.; Zhou, R. Techno-economic evaluation of a community-based hybrid renewable energy system considering site-specific nature. *Energy Convers. Manag.* **2018**, *171*, 1737–1748. [CrossRef]
24. Jones, C.R.; Eiser, J.R. Understanding 'local' opposition to wind development in the UK: How big is a backyard? *Energy Policy* **2010**, *38*, 3106–3117. [CrossRef]
25. Dimitropoulos, A.; Kontoleon, A. Assessing the determinants of local acceptability of wind-farm investment: A choice experiment in the Greek Aegean Islands. *Energy Policy* **2009**, *37*, 1842–1854. [CrossRef]
26. Kalkbrenner, B.J.; Roosen, J. Citizens' willingness to participate in local renewable energy projects: The role of community and trust in Germany. *Energy Res. Soc. Sci.* **2016**, *13*, 60–70. [CrossRef]
27. Devine-Wright, P. Local aspects of UK renewable energy development: Exploring public beliefs and policy implications. *Local Environ.* **2005**, *10*, 57–69. [CrossRef]
28. Salm, S.; Hille, S.L.; Wüstenhagen, R. What are retail investors' risk-return preferences towards renewable energy projects? A choice experiment in Germany. *Energy Policy* **2016**, *97*, 310–320. [CrossRef]
29. Masini, A.; Menichetti, E. The impact of behavioral factors in the renewable energy investment decision making process: Conceptual framework and empirical findings. *Energy Policy* **2011**, *40*, 28–38. [CrossRef]
30. Kosenius, A.K.; Ollikainen, M. Valuation of environmental and societal trade-offs of renewable energy sources. *Energy Policy* **2013**, *62*, 1148–1156. [CrossRef]
31. Scarpa, R.; Willis, K. Willingness-to-pay for renewable energy: Primary and discretionary choice of British households' for micro-generation technologies. *Energy Econ.* **2010**, *32*, 129–136. [CrossRef]
32. Train, K. *Discrete Choice Methods with Simulation*, 2nd ed.; Cambridge University Press: Cambridge, UK, 2003.
33. Hanley, N.; Wright, R.E. Using choice experiments to value the environment. *Environ. Resour. Econ.* **1998**, *11*, 413–428. [CrossRef]
34. Yang, H.J.; Lim, S.Y.; Yoo, S.H. The environmental costs of photovoltaic power plants in South Korea: A choice experiment study. *Sustainability* **2017**, *9*, 1773. [CrossRef]
35. Haaijer, R.; Wedel, M. Conjoint choice experiments: General characteristics and alternative model specification. In *Conjoint Measurement: Methods and Applications*, 3rd ed.; Gustafsson, A., Herrmann, A., Huber, F., Eds.; Springer: Berlin, Germany, 2003; pp. 371–412.
36. KEEI (Korea Energy Economics Institute). *2019 Yearbook of Energy Statistics*; KEEI: Ulsan, Korea, 2019. (In Korean)
37. MOTIE. *The Renewable Energy 3020 Implementation Plan*; MOTIE: Sejong, Korea, 2017. (In Korean)
38. Im, H.; Yun, S.J. Analysis of the policy process of the separation distance regulations of local governments concerning the location conflicts of photovoltaics facilities. *New Renew. Energy* **2019**, *15*, 61–73. (In Korean) [CrossRef]
39. Park, Y.M.; Choung, T.R.; Son, J.H. Study on noise and low frequency noise generated by wind power plant (wind farm). *J. Environ. Impact Assess.* **2011**, *20*, 425–434. (In Korean)
40. Haugen, K.M.B. *International Review of Policies and Recommendations for Wind Turbine Setbacks from Residences: Setbacks, Noise, Shadow Flicker, and Other Concerns*; Minnesota Department of Commerce: Saint Paul, MN, USA, 2011.
41. MOTIE. *Announcement of Renewable Energy 3020 Implementation and Countermeasures to Resolve Side Effects of Solar Photovoltaic and Wind Power*; MOTIE: Sejong, Korea, 2018; (press release, In Korean).
42. Langer, K.; Decker, T.; Menrad, K. Public participation in wind energy projects located in Germany: Which form of participation is the key to acceptance? *Renew. Energy* **2017**, *112*, 63–73. [CrossRef]

43. Lienhoop, N. Acceptance of wind energy and the role of financial and procedural participation: An investigation with focus groups and choice experiments. *Energy Policy* **2018**, *118*, 97–105. [CrossRef]
44. Liu, L.; Bouman, T.; Perlaviciute, G.; Steg, L. Effects of trust and public participation on acceptability of renewable energy projects in the Netherlands and China. *Energy Res. Soc. Sci.* **2019**, *53*, 137–144. [CrossRef]
45. Suškevičs, M.; Eiter, S.; Martinat, S.; Stober, D.; Vollmer, E.; de Boer, C.L.; Buchecher, M. Regional variation in public acceptance of wind energy development in Europe: What are the roles of planning procedures and participation? *Land Use Pol.* **2019**, *81*, 311–323. [CrossRef]
46. Bauwens, T. Explaining the diversity of motivations behind community renewable energy. *Energy Policy* **2016**, *93*, 278–290. [CrossRef]
47. Hammami, S.M.; Chtourou, S.; Triki, A. Identifying the determinants of community acceptance of renewable energy technologies: The case study of a wind energy project from Tunisia. *Renew. Sustain. Energy Rev.* **2016**, *54*, 151–160. [CrossRef]
48. Langer, K.; Decker, T.; Roosen, J.; Menrad, K. A qualitative analysis to understand the acceptance of wind energy in Bavaria. *Renew. Sustain. Energy Rev.* **2016**, *64*, 248–259. [CrossRef]
49. Bateman, I.J.; Carson, R.T.; Day, B.; Hanemann, M.; Hanley, N.; Hett, T.; Jones-Lee, M.; Loomes, G.; Mourato, S.; Özdemiroglu, E.; et al. *Economic Valuation with Stated Preference Techniques: A Manual*; Edward Elgar Publishing Ltd.: Cheltenham, UK, 2002.
50. KEEI. *Research on Residents Participatory New Renewable Energy Power Plant Promotion Plan*; MOTIE: Sejong, Korea, 2014. (In Korean)
51. Korea Energy Agency. *Residents Participatory New and Renewable Power Generation Project Incentive Plan*; Korea Energy Agency: Yongin, Korea, 2016. (In Korean)
52. Ouedraogo, B. Household energy preferences for cooking in urban Ouagadougou, Burkina Faso. *Energy Policy* **2006**, *34*, 3787–3795. [CrossRef]
53. Rao, M.N.; Reddy, B.S. Variations in energy use by Indian households: An analysis of micro level data. *Energy* **2007**, *32*, 143–153.
54. McFadden, D. Conditional logit analysis of qualitative choice behavior. In *Frontiers of Econometrics*, 1st ed.; Zarembka, P., Ed.; Academic Press: New York, NY, USA, 1974; pp. 142–150.
55. Guo, S.; Fraser, M.W. *Propensity Score Analysis: Statistical Methods and Applications*, 2nd ed.; SAGE: Thousands Oaks, CA, USA, 2014.
56. Borchers, A.M.; Duke, J.M.; Parsons, G.R. Does willingness to pay for green energy differ by source? *Energy Policy* **2007**, *35*, 3327–3334. [CrossRef]
57. Ma, C.; Rogers, A.A.; Kragt, M.E.; Zhang, F.; Polyakov, M.; Gibson, F.; Chalak, M.; Pandit, R.; Tapsuwan, S. Consumers' willingness to pay for renewable energy: A meta-regression analysis. *Resour. Energy Econ.* **2015**, *42*, 93–109. [CrossRef]
58. Ribeiro, F.; Ferreira, P.; Araújo, M.; Braga, A.C. Public opinion on renewable energy technologies in Portugal. *Energy* **2014**, *69*, 39–50. [CrossRef]
59. Lee, H.J.; Huh, S.Y.; Woo, J.; Lee, C.Y. A comparative study on acceptance of public and local residents for renewable energy projects: Focused on solar, wind, and biomass. *Innov. Stud.* **2020**, *15*, 29–61. (In Korean)
60. Burningham, K.; Barnett, J.; Walker, G. An array of deficits: Unpacking NIMBY discourses in wind energy developers' conceptualizations of their local opponents. *Soc. Nat. Resour.* **2015**, *28*, 246–260. [CrossRef]
61. Ek, K. Public and private attitudes towards "green" electricity: The case of Swedish wind power. *Energy Policy* **2005**, *33*, 1677–1689. [CrossRef]
62. Koch, J.; Christ, O. Household participation in an urban photovoltaic project in Switzerland: Exploration of triggers and barriers. *Sust. Cities Soc.* **2018**, *37*, 420–426. [CrossRef]
63. Liu, B.; Hu, Y.; Wang, A.; Yu, Z.; Yu, J.; Wu, X. Critical factors of effective public participation in sustainable energy projects. *J. Manag. Eng.* **2018**, *34*, 04018029. [CrossRef]

 energies

Article

Socio-Economic Effect on ICT-Based Persuasive Interventions Towards Energy Efficiency in Tertiary Buildings

Diego Casado-Mansilla [1,*], Apostolos C. Tsolakis [2], Cruz E. Borges [1], Oihane Kamara-Esteban [1], Stelios Krinidis [2], Jose Manuel Avila [3], Dimitrios Tzovaras [2] and Diego López-de-Ipiña [1]

1 DeustoTech—Facultad Ingeniería, Universidad de Deusto, Avda. Universidades, 24, 48007 Bilbao, Spain; cruz.borges@deusto.es (C.E.B.); oihane.esteban@deusto.es (O.K.-E.); dipina@deusto.es (D.L.-d.-I.)

2 Information Technologies Institute, Centre for Research and Technology, Hellas, 57001 Thessaloniki, Greece; tsolakis@iti.gr (A.C.T.); krinidis@iti.gr (S.K.); dimitrios.tzovaras@iti.gr (D.T.)

3 Wellness TechGroup, Calle Gregor J. Mendel, 6, 41092 Sevilla, Spain; jmavila@wellnesstg.com

* Correspondence: dcasado@deusto.es

Received: 26 February 2020; Accepted: 30 March 2020; Published: 3 April 2020

Abstract: Occupants of tertiary environments rarely care about their energy consumption. This fact is even more accentuated in cases of buildings of public use. Such unawareness has been identified by many scholars as one of the main untapped opportunities with high energy saving potential in terms of cost-effectiveness. Towards that direction, there have been numerous studies exploring energy-related behaviour and the impact that our daily actions have on energy efficiency, demand response and flexibility of power systems. Nevertheless, there are still certain aspects that remain controversial and unidentified, especially in terms of socio-economic characteristics of the occupants with regards to bespoke tailored motivational and awareness-based campaigns. The presented work introduces a two-step survey, publicly available through Zenodo repository that covers social, economic, behavioural and demographic factors. The survey analysis aims to fully depict the drivers that affect occupant energy-related behaviour at tertiary buildings and the barriers which may hinder green actions. Moreover, the survey reports evidence on respondents' self-assessment of fifteen known principles of persuasion intended to motivate them to behave pro-environmentally. The outcomes from the self-assessment help to shed light on understanding which of the Persuasive Principles may work better to nudge different user profiles towards doing greener actions at workplace. This study was conducted in four EU countries, six different cities and seven buildings, reaching more than three-hundred-and-fifty people. Specifically, a questionnaire was delivered before (PRE) and after (POST) a recommendation-based intervention towards pro-environmental behaviour through Information and Communication Technologies (ICT). The findings from the PRE-pilot stage were used to refine the POST-pilot survey (e.g., we removed some questions that did not add value to one or several research questions or dismissed the assessment of Persuasive Principles (PPs) which were of low value to respondents in the pre-pilot survey). Both surveys validate "Cause and Effect", "Conditioning" and "Self-monitoring" as the top PPs for affecting energy-related behaviour in a workplace context. Among other results, the descriptive and prescriptive analysis reveals the association effects of specific barriers, pro-environmental intentions and confidence in technology on forming new pro-environmental behaviour. The results of this study intend to set the foundations for future interventions based on persuasion through ICT to reduce unnecessary energy consumption. Among all types of tertiary buildings, we emphasise on the validity of the results provided for buildings of public use.

Keywords: occupant behaviour; socio-economic profile; survey; energy efficiency; persuasion; intervention; pro-environmental behaviour change; workplace

1. Introduction

Energy efficiency and, even more so, energy flexibility approaches related to demand response are inseparably connected with energy-related occupant behaviour. For several years, scientists have tried to tackle the challenge of deconstructing occupant activities and choices in the home and work environment, introducing various behavioural models and adapting previous ones from other scientific fields into the energy sector [1,2]. However, scholars and relevant stakeholders do not always succeed in the study of energy-specific related behaviour. This fact is emphasised in a working environment which seems to be more challenging; working with different user-profiles with different values, beliefs and norms interacting in the same place [3]. Even in recent reviews [4,5], where the high energy-saving potential of occupant behaviour is identified (up to 30% for tertiary buildings), there are still missing aspects that elude the scientific community. These are, but not limited to, systematic frameworks for understanding occupant behaviour that go beyond individual buildings, cultures and geographical boundaries, as well as a more elaborate understanding of persuasion or incentivisation mechanisms for improving more energy efficiency practices. The body of literature reviewed in this article employed questionnaires, interviews or surveys to shed light on those open challenges. These surveys cover a vast area of different parameters but present limitations when trying to accurately pinpoint the factors that affect the behaviour of the occupants towards improving energy performance at working environments.

Energy-related behaviour has been studied for quite a few decades. Over the years, certain theories have been proven to present better results when studying pro-environmental behaviour, with the most well known being Ajzen's planned behaviour theory [6], Hines et al.'s model of responsible environmental behaviour [7] and Stern's value–belief–norm (VBN) theory [8]. As technology improved, behavioural energy approaches have become more sophisticated, integrating the ongoing interaction from multiple drivers/factors [4,5,9]. Yet, despite the efforts on delivering a more integrated scheme, they seem to lack certain aspects, as denoted in the literature. This is especially highlighted by a recent review on energy-related occupant behaviour surveys [10], where in most of the 33 projects that were analysed, essential issues in social science were disregarded, and many other vital aspects of human behaviour were not measured or considered. Some interesting limitations are the lack of in-depth analysis and understanding of (i) the effect that different cultures, countries and climates introduce to occupant behaviour; (ii) the users' actions for restoring comfort conditions; and (iii) the effect that group/collective behaviour may have to energy-related aspects.

If modelling and understanding the user behaviour within buildings is crucial to detect energy leaks and the barriers that prevent them from behaving more energy-efficiently, the same relevance has to be provided to campaigns and interventions to motivate and engage users into green practices. Towards that direction, a new concept in behavioural science has been introduced: the nudge theory (or nudging), which eventually aims to influence the motives, incentives and decision-making of groups and individuals [11]. Investing in this concept, numerous variations have been presented either as tools for deeply understanding factors/drivers that lead end-users to act in a certain way or as a means to persuade and change the overall behaviour. Furthermore, many persuasion mechanisms have been proposed to influence energy-related behaviour towards more energy-efficient actions. Persuasion for sustainability has its roots in the application of Fogg's framework [12] for "computers as persuaders" to the topic of pro-environmental sustainability. Specifically, in interventions based on Information and Communication Technologies (ICT), the use of Fogg's or Cialdini's [13] theories to provide nudges adopted the term "eco-feedback" [14], which, in short, informs users about their actions and make them reflect over resource waste through ambient feedback. Whereas these are widely provided on health and energy efficiency intervention, persuasion and nudging are not exempt from controversy. There are recent works which lower the impact they have [15,16] or voices that raised their concern about the feasibility of persuasion to maintain the target behaviour in the long-term [17,18]. In short, going beyond understanding energy-related behaviour, when trying to persuade end-users to alter their energy-expensive behaviour, the proper persuasion tools that have the desired effect to occupants—even more in cases of tertiary buildings—still intrigue the scientific community.

In order to link the two presented challenges, this work presents a survey that takes into account both socio-economic factors and energy-related aspects to better understand the human drivers that could be integrated on a new ICT-based persuasion-based engine for improving occupant behaviour at an individual's working environment. The survey responses help identify which profiles and principles are more likely to instil behavioural change towards a more energy-efficient practice at work by building a socio-economic profile for each of the occupants of the workplace. One of the main particularities of the survey is that it was conducted in seven different buildings across six different cities in four EU countries and reaching more than three-hundred-and-fifty people (350). Therefore, the analysis of the dependencies between factors and principles takes a broader perspective considering the application of the questionnaire to people with a very different background and a heterogeneous socio-demographic profile. Further refinement of the survey was also elaborated after the assessment of a behavioural change intervention, highlighting the factors that were identified in the beginning but did not present significant results after the first iteration of responses.

The presented work is structured as follows. Section 2 presents the GreenSoul project and provides the basis of the intervention we carried out, followed by Section 3 that explains the design of the survey performed. Section 4 delivers the statistical results, and Section 5 provided a thorough discussion on them along with the limitations of the study. Finally, Section 6 concludes the manuscript containing a final outlook of research.

2. The GreenSoul Project

In an effort to achieve higher energy efficiency in public buildings, the GreenSoul (GS) project [19], an H2020 research and innovation action, designed, implemented and tested a system for saving energy in tertiary buildings allowing to produce energy savings up to 25% in the best scenario [20]. GS is based on monitoring devices of personal (e.g., laptops or monitors) and shared equipment (e.g., lighting, HVAC or appliances) and mechanisms to give personalised usage recommendations and nudges, taking into consideration the characteristics of the users and adapting the feedback to their changes (the feedback is different depending if the user should form, enhance or maintain an eco-behaviour). One of the hypotheses of the project is that each user-profile has an optimal way to receive feedback on their energy consumption and good practices to increase energy performance. Therefore, profiles are identified through online questionnaires and the system makes use of this information to create a tailored socio-economic model of the user to incentive him/her to save energy through innovative ICT elements. By investing in end-user engagement, rather than sophisticated automation or Building Management Systems (BMS), the cost is reduced and the results shall be prolonged. To encourage users to reduce energy consumption, an additional effort has been included to persuade them to be energy efficient (the impact of daily habits on the energy consumed at work is mainly reflected in the electricity bill, which never reaches the employees). To this end (a) the behaviour of people related to energy efficiency at work is modelled; (b) a socio-economic characterisation and segmentation is carried out using statistical analyses; (c) energy monitoring hardware was deployed in each of the six pilots; and finally, (d) the effectiveness of different ICT-based persuasion strategies provided through different interactive channels was tested in each of the identified population segments.

2.1. The GreenSoul Intervention

To test the effectiveness of the overall GreenSoul system, a Randomised Control Trial (RCT) was carried out in seven pilot buildings across Europe involving more than three-hundred-and-fifty people (350). Four different treatments combining three different persuasion principles (i.e., cause–effect, self-monitoring and conditioning) through ICT were deployed. These treatments were delivered using different feedback channels: a custom-based interactive coaster that provided visual information about energy consumption (self-monitoring), a gamified mobile app with some automation features (conditioning), a series of analogue signage in the form of post-its and posters with "green messages" (cause–effect), which can be considered as the control-treatment, and all of the three previous treatments altogether. Figure 1 shows a picture of them.

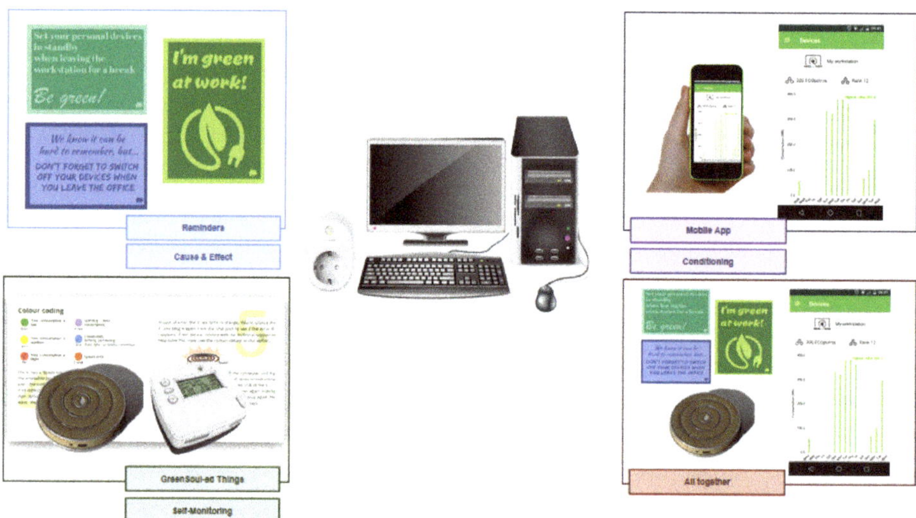

Figure 1. The GreenSoul persuasion treatments.

We prepared the intervention on the basis of the following hypothesis; individualised persuasion messages based on personal energy consumption produce a high engagement, but the potential impact is low compared to the energy consumption of the whole building. However, if an intervention started addressing directly the collective, its potential impact on energy-savings is higher, but end-users might not be motivated to pursue green-actions (see Figure 2). To overcome this problem, the RCT was divided into two phases: individual and collective. During the individual phase, the primary objective was to foster the awareness and motivation of the participants in energy efficiency practices. Therefore, we only provided individual information regarding their performance with devices and appliances under their own control. In the second phase (with all participants already engaged), we gave persuasive cues about how to reduce the energy consumption of electricity-powered devices not directly attached to the individual but more related to equipment of shared use (e.g., lighting, HVAC or common appliances). In workplaces, there is usually no one in charge of them and the diffusion of responsibility and social loafing appear as the primary factors for energy waste [21]. Indeed, according to Whittle and Jones [22], "Both theories indicate that in group situations people are generally less likely to take action than when alone because, for instance, they do not assume personal responsibility for taking action (anticipating that someone else will do so) or because they believe that others will not put in the effort, so do not put in (as much) effort themselves."

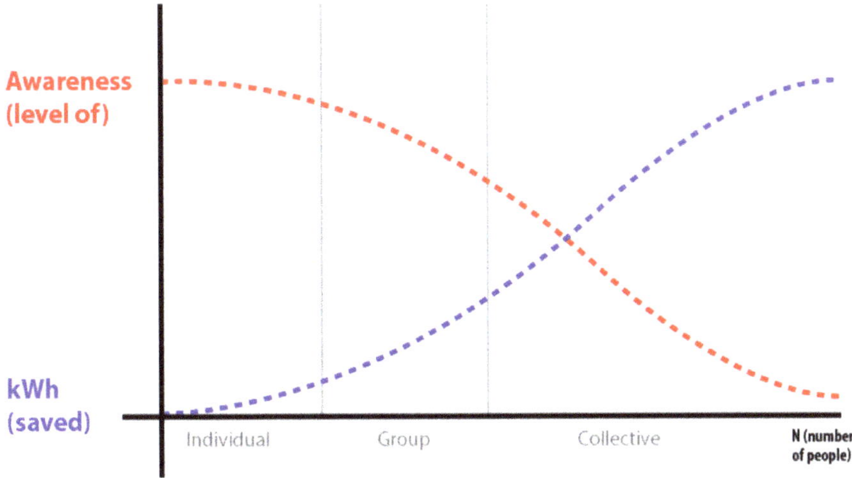

Figure 2. The GreenSoul phases: Individual where motivation is high but the effect on energy efficiency and reduction is low. Collective, where the motivation of users to embark on campaigns might be low, but the potential of the energy savings is high.

2.2. Evaluation Procedure

We triangulated the overall GreenSoul solution from three different qualitative and quantitative sources: (1) PRE-POST validated surveys to assess energy awareness, motivations to change the behaviour and main obstacles that hinder the adoption of energy practices in the workplace; (2) the energy consumption per user, per treatment and per building along the whole study; (3) focus groups throughout the whole experimental phases to understand user motivations at each time, interventions pitfalls and other relevant matters. In this article we put the focus on the questionnaires to provide a thorough analysis of the user profiles and their specific motivations and barriers to behave pro-environmentally at the workplace.

3. Survey Design and Delivery

The survey presented in this manuscript was developed in two stages: the PRE and POST pilot interventions. Starting with a state-of-the-art analysis, validated instruments and acquired knowledge from previous experiences, the initial survey was constructed, before applying any interventions to the buildings, consisting of 36 groups of questions divided into four sections [23]. The final outcome provides a complete demographic and socio-economic outline of the respondents including also a section for their self-rate preferences in terms of what persuasion strategies they would like to be applied in an energy-related intervention in the workplace. The PRE-pilot full four sections are briefly explained in the following sections. A more elaborated description is provided in [24].

3.1. Social, Economic and Demographic Traits

Questions related to the personal information of the users are essential towards being able to identify general groups and profiles that may be more keen to change based on the suggestions provided. We subdivide such information into those factors related to the workplace and those which are independent.

1. *Non-Dependent on Work*: Common questions that cover demographic (static) aspects such as age, gender, education, country and city. Additionally, this group includes questions related also to dynamic aspects such as confidence in technology, susceptibility to persuasion as well as attitudinal and intentional profiles.
2. *Dependent on Work*: Questions directly influenced by the organisation where the user is employed, consisting once again of static information such as the type of employment, the position, work culture, etc. and dynamic energy-related factors such as the main barriers to be energy efficient, the willingness to join an initiative to reduce energy consumption, and more.
3. *Energy-related actions at work*: An ancillary profile was constructed through questions that cover energy-related social habits such as the use of the HVAC, the set-point established in summer and winter, the use of natural and artificial lighting, printing habits or the power mode set in users' equipment.
4. *Ranking of Persuasion Strategies*: Users were asked to self-rate (one to five likert scale) 21 persuasive strategies related to enhancing energy efficiency practices in shared spaces at working environments. The strategies were designed based on fifteen persuasion principles developed by experts [12,13,25]. Table 1 presents the Persuasion Principles, Table 2 introduces the Persuasion Strategies (numbered from v2 to v22 as this is the coding they have on Zenodo's dataset), and in Table 3 these two are mapped together.

In order to evaluate and validate the interventions of the GreenSoul project, a second survey was constructed after the completion of the pilot execution period. Given the findings originating from the analysis of the first survey, a lot of socio-economic questions were omitted in the second version (e.g., country, having children or not, current employment status, are you satisfied with your thermal comfort at workplace, etc.), as they were not found to have significant influence to any of the Persuasive Principles under study. Besides, we removed the five least ranked principles of persuasion by initial respondents to only provide again those that might have higher impact influencing people from their point of view. Finally, we added ten questions aiming to identify the perspective of end-users regarding the project and its effect.

The two surveys designed for the PRE-pilot and POST-pilot assessments along with their raw results can be found on Zenodo in [23] and [26], respectively.

Table 1. Explanation and Rationale of 15 Persuasion Principles.

#	Persuasion Principle	Description
P1	Authority	People will tend to obey authority figures, even if they are asked to perform objectionable acts. People want to follow the lead of real experts.
P2	Cause and effect	This principle can effectively persuade people to change their attitudes or behaviour by enabling them to observe a direct link between cause and effects of daily actions.
P3	Conditioning	A behavioural process whereby a response becomes more frequent or more predictable in a given environment as a result of reinforcement, with reinforcement typically being a stimulus or reward for a desired response. Computerised systems use principles of operant conditioning to change behaviours. To use operant conditioning is basically to provide reinforcement (positive or negative) to the user.
P4	Cooperation Liking	A system can motivate users to adopt a target attitude or behaviour by leveraging human beings' natural drive to cooperate. Further, People prefer to say "yes" to those they know and like. People are also more likely to favour those who are physically attractive, similar to themselves, or who give them compliments. Even something as "random" as having the same name as your prospects can increase your chances of making a sale (e.g., Coke).
P5	Tailoring Personalisation	A system that offers personalised content or services has a greater capability for persuasion. Further, information provided by the system will be more persuasive if it is tailored to the potential needs, interests, personality, usage context, or other factors relevant to a user group. It provides information that is specific to the individual to better enable a certain behaviour.
P6	Physical attractiveness	It plays an essential role in making the application easy to understand and more likeable.
P7	Praise	By offering praise, a system can make users more open to persuasion.
P8	Real-world feel verifiability	A system that highlights people or organisation behind its content or services will have more credibility. System should provide information of the organisation and/or actual people behind its content and services. Moreover, credibility perceptions will be enhanced if a system makes it easy to verify the accuracy of site content via outside sources.
P9	Reciprocity	Reciprocation recognises that people feel indebted to those who do something for them or give them a gift. People tend to return a favour.
P10	Reduction	A system that reduces complex behaviour into simple tasks helps users perform the target behaviour, and it may increase the benefit/cost ratio of a behaviour.
P11	Self-monitoring	A system that keeps track of one's own performance or status supports the user in achieving goals. The goal is to allow people to monitor themselves to modify their behaviour to achieve a predetermined objective or outcome.
P12	Similarity	People are more readily persuaded through systems that remind them of themselves in some meaningful way. We are more persuaded by people we think are similar to us (personality, interests, etc).
P13	Social proof	A psychological phenomenon where people assume the actions of others in an attempt to reflect correct behaviour for a given situation. It is the concept that people will conform to the actions of others under the assumption that those actions are reflective of the correct behaviour.
P14	Social Recognition	By offering public recognition for an individual or group, a system can increase the likelihood that a person/group will adopt a target behaviour.
P15	Suggestion	Systems offering fitting suggestions will have greater persuasive powers. People are more likely to engage in an activity when it is closely related to what they are currently doing.

Table 2. Description of the 21 Persuasion Strategies Identified.

#	Persuasion Strategy
v2	Public (social) recognition of your contribution to energy savings is provided
v3	Receive personal praise (privately) for your contribution to energy savings
v4	The support of the majority of your peers to improve energy efficient behaviour
v5	Receive energy related information in a simple and aesthetically appealing way
v6	Receiving perks such as flexible working hours, skipping certain tasks, etc., as a reward for improving your energy performance
v7	You and your team receive recognition for collectively achieving energy savings
v8	You receive information about the people (e.g., engineers, vendors, etc.) behind the instruments and equipment which allows you to collect energy-related data
v9	You are assisted in setting, meeting and reviewing your own personal energy saving goals
v10	Your (top) managers are also committed to save energy
v11	You can monitor & track your own energy performance in real-time
v12	The overall energy saving goals are broken down into smaller easily achievable
v13	The feasibility of the proposed energy savings has been verified in other buildings similar to your workplace
v14	Energy related information is tailored to you and you are able to self-configure some parameters (e.g., data provided, frequency, etc.) according to your preferences
v15	Information on the actual effect that your (potential) actions may have upon the energy consumption
v16	Comparative assessment of your actual energy performance compared to benchmarks/ good practices
v17	Comparative assessment of your energy saving performance with the respective performance of your peers (e.g., colleagues, other visitors, etc.)
v18	Historical comparison of your energy performance and/or consumption
v19	Tips or suggestions on the energy saving practice of the day/week
v20	Progress, tips and lessons learned on specific energy saving actions performed by other users that are similar to me
v21	Advice and quotes from energy experts (including external energy consultants, energy researchers, energy agencies, etc.)
v22	Links to data about how energy consumption is monitored and (potential) energy savings assessed

Table 3. Mapping of persuasive strategies to the persuasion principles.

Persuasion Principle	Persuasion Strategy
Authority (P1)	v10, v21
Cause and effect (P2)	v15
Conditioning (P3)	v6
Cooperation & Liking (P4)	v4
Tailoring & Personalization (P5)	v9, v14
Physical attractiveness (P6)	v5
Praise (P7)	v3
Verifiability & Real-world feel (P8)	v8, v13, v22
Reciprocity (P9)	v7
Reduction (P10)	v12
Self-monitoring (P11)	v11, v16, v18
Similarity (P12)	v20
Social proof (P13)	v20, v10
Social Recognition (P14)	v2, v7
Suggestion (P15)	v19

3.2. Survey Setup

An online survey was delivered in seven official tertiary pilot buildings in the following cities; Bilbao—Spain, Cambridge—UK, Sussex—UK, Pilea-Hortiatis—Greece, Seville—Spain, Thessaloniki—Greece and WEIZ—Austria. The survey was delivered in two stages, prior the intervention of the GreenSoul framework (Pre-pilot—between May and July 2017) and after it (Post-pilot November 2019) (A description of the pilots is provided in the Appendix A of this manuscript). For the GreenSoul pilot execution, only five out of the seven buildings were used, hence allowing the existence of control groups (in our case ECOLUTION (Sussex) and CERTH (Thessaloniki)).

In total, for the Pre-Pilot survey three-hundred-and-twenty-three (323) responses to the questionnaires were collected. After conducting a data cleaning process (i.e., removing uncompleted questionnaires and outliers) three-hundred-and-three (303) samples remained to be analysed. Of these responses, eighty-four (84) correspond to people working in the Greek sites (PILEA & CERTH), eighty-three (83) were those working in Spain sites (DEUSTO & SEVILLE), eighteen (18) were those working in Austria (WEIZ) and, most of the answers, one-hundred-eighteen (118) respondents working in the UK (ALLIA & ECOLUTION). Accordingly, for the Post-Pilot survey after two years of intervention, one hundred and five (105) responses were retrieved from the pilots. Of these responses, thirty-seven (37) correspond to people working in the Greek sites, forty-three (43) were those working in Spanish sites, twelve (12) were those working in Austria and thirteen (13) respondents working in the UK. Table 4 summaries the number of completed questionnaires collected per pilot site per survey trial period. The surveys were deployed online using Google Forms, and they were distributed through email to all sites in their national language. The participation was voluntary and a pilot responsible provided reminders during the survey period to ensure an adequate participation rate per site.

Table 4. Number of responses per GreenSoul cities

	Bilbao	Cambridge	Sussex	Pilea	Seville	Weiz	Thessaloniki
Pre-Pilot	53	58	60	26	30	18	58
Post-Pilot	28	8	5	19	15	12	18

4. Results

The analysis of the responses collected has been performed in both a descriptive and a prescriptive manner. In the former, the analysis aims to provide some basic descriptive statistics and conclusions over the data collected. Regarding the socio-economic and demographics, we can conclude which are the prominent profiles among different pilot buildings, whether we find gender or technical divide in the population or if our sample is willing to join energy initiatives at work or they are reluctant. In relation to the persuasive part, we can conclude which are the most and least rated persuasive strategies and the principles behind them. The prescriptive analysis is devoted to identifying socio-economic factors that are potentially important in determining certain persuasion principles to be used or which profiles would be more prone to elicit changes in pro-environmental behaviour according to persuasive strategies behind these principles.

4.1. PRE/POST Preprocessing

Before analysing the results from the two surveys, a preprocessing of the received data was necessary. For the first survey, given the size of the overall questionnaire, a dimensionality reduction of the overall system was decided. All the factors related to habits in the workplace (i.e., Ancillary profile) were excluded, as they are tightly dependent on the workplace where employees were at the moment of answering the questionnaire. Therefore, we finally analysed the socio-economic profile with the following variables; Age, Gender, Education, Country, City, Employment, Position, Work culture, Sharing (whether people share the room at work with others or not), Work activity, Profile PST (Pinball, shortcut, thoughtful) [27], Intentions (to behave pro-environmentally), Confidence (in technology), Organisation energy (if employees perceive their organization as green or not), Barriers (to behave pro-environmentally in the workplace), Consensus (difficulty or ease of reaching consensus with peers), Influences (if people tend to influence others or not), Susceptibility (of people to persuasion), Initiative to join (environmental campaigns), desired Frequency to receive feedback and Response provided to green signage.

Regarding the persuasive strategies, per user we found the sub-set of Persuasive Principles that were rated with the highest rank, and we distributed one score among them. Thus, if an employee rated N principles, we assigned $1/N$ points to each principle. After finalising the process, we came up with a dataset of principles of persuasion with a score in only those which were voted with the top rank. Finally, questions that were not answered were not included in the analysis.

4.2. Descriptive Analysis: PRE vs. POST

As the PRE-pilot descriptive analysis has been described thoroughly in [24], the comparative analysis of the two surveys is presented next highlighting the most interesting aspects identified in both timeframes.

In the POST-pilot survey analysis a slight shift is observed towards male (61.9%) over female (37.1%) respondents, whereas the PRE-pilot was more balanced with 48.1% women and 51.9% men. The principal age group remained at the range of 21 to 40 (49.5%), whereas the percentage of participants holding a master's degree has elevated by 10% (POST: 41.0% over PRE: 31.7%), and the respondents that have only finished high school has slightly diminished (POST: 4.8% over PRE: 5.1%). The position has been divided in more groups with 73.33% working as employees (compared to the previous PRE: 88.5%), 8.57% as higher positions (principal researchers, head of unit or boss), and 16.19% as administrative staff. Following the same analysis as the PRE-pilot questionnaire, the profiles identified, according to Lockton's model of the user [27], were in the following order; Thoughtful (POST: 44.21% instead of PRE: 54.3%), followed by Pinball (POST: 15.79% instead of PRE: 11%) and Shortcut (POST: 14.74% instead of PRE: 14.3%). Therefore, we observed that the differences among the two snapshots in PRE and POST do not vary significantly in socio-economic terms. Thus, for this study, we assume that the two samples could be considered as comparable units for evaluation purposes. In the limitations section at the end of the manuscript we elaborate on this issue.

In terms of persuasion principles, even though fifteen principles were originally identified by experts, most of them were not found to be important enough for end-users to receive significant attention from them. As a result, in the POST-pilot survey, only the top ten principles that were highly rated in the PRE-pilot were included. As can be observed in Figure 3, there were quite a few differences when rating these by the end-users prior and after the interventions. However, we can observe that Cause and Effect (P2), Self-monitoring (P11) and Conditioning principle (P3) were top rated in both surveys.

From the answers received about employees' intentions to behave in favour of the environment using as instrument the validated questionnaire provided by [28], an interesting increase has been observed in the POST-pilot survey in the Action stage (PRE: 47.84% vs. POST: 56.38%), followed by a reduction in the Contemplation (PRE: 48.17% vs. POST: 38.30%). These evaluated stages are alike to those extracted from Transtheoretical Model of behaviour change (TTM) [29], which can be observed in Figure 4.

According to the provided results, it seems that end-users have shifted from hesitation to real determination to act in a more energy efficient manner after the GS intervention was provided. Statistically speaking, we found an important difference in Pre-Contemplation stage between the responses of PRE and POST pilots questionnaires when treating the dataset as a whole (i.e., pooled data). Wilcoxon Signed-Ranks Test indicated that the median pre-test ranks were statistically significantly higher than the median post-test ($W = 1, Z = 4.5399, p < 0.05, r = 0.2225$). Thus, the average mean of people in Pre-Contemplation stage before the treatments was greater than the computed average at the end of the interventions. Thus, the overall employees participating in GreenSoul project seem to have reduced their stage of Pre-contemplation in favour to higher stages, where the willingness to do action in favour of the environment increased. This observation has a medium effect size according to Cohen's criteria [30].

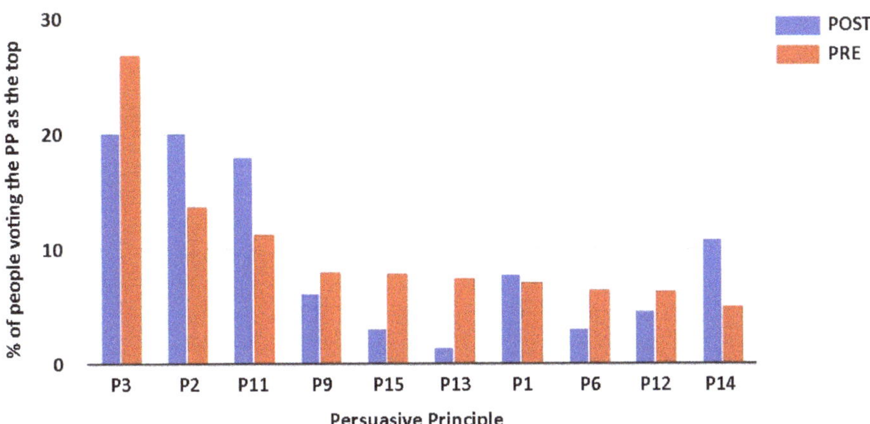

Figure 3. PRE vs. POST-pilot: Ranking of the top ten dominant persuasion principles as voted by the end-users as a top principle.

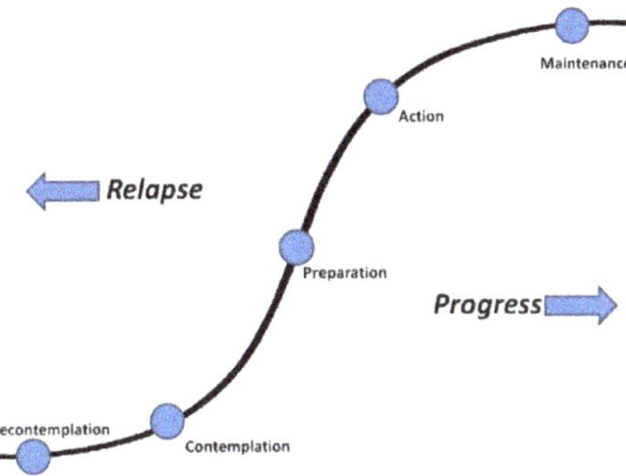

Figure 4. The stages of change according to the transtheoretical model to evaluate enhancement or relapse of behaviour change.

Descriptive results so far covered the respondents' opinions using a pooled sample approach. When examining pilot by pilot (recall that a description of the pilots is provided in Appendix A), some interesting outcomes were observed.

4.2.1. SEVILLE

A contingency analysis for the barriers using Fisher's test [31] showed that people in this pilot increased their overall intentions in favour of the environment significantly (p-value < 0.02). Specifically, we can observe in Figure 5 that the percentage of people in the Contemplation stage (2) was reduced and increased in Action stage (3) in the POST-pilot results. Pre-contempation (1) in PRE and POST was similar.

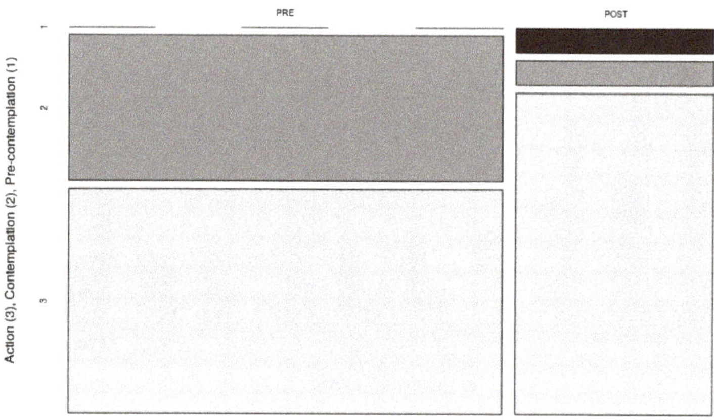

Figure 5. Overall pro-environmental Intentions in Seville (PRE vs. POST-pilot). We can observe the reduction of people in contemplation stage and the relevant increase of people reporting to be in Action.

4.2.2. DEUSTO

We observed an important difference in Pre-contemplation stage between the responses of PRE and POST pilot questionnaires in this Spanish pilot ($W = 1, Z = 5.1428, p < 0.05, r = 0.5749$). According to the large effect size observed ($r = 0.5749$) with regards to Cohen's criteria [30], the results might suggest that this pilot contributed importantly to the effect observed above in the pre–post analysis when examining the pooled samples approach. Furthermore, the tests in Contemplation and Action stages confirmed that not only people in this pilot scored lower on Pre-contemplation, but they also increased their stages of change towards more conscious eco-behaviour. Contemplation ($W = 1, Z = -2.0434, p < 0.05, r = 0.228$) and Action ($W = 1, Z = -2.4688, p < 0.05, r = 0.5749$).

4.2.3. WEIZ

According to the results, the medians of PRE and POST regarding Pre-contemplation stage in Weiz were 0.2 and -1.47, respectively. A Wilcoxon Signed-rank test shows that there is a significant effect of Group WEIZ with the Pre-contemplation stage of the TTM ($W = 1, Z = 1.9598, p < 0.05, r = 0.3464$). That means that the respondents change their Pre-contemplation stage towards higher stages of consciousness. Interestingly, Contemplation also differed statistically ($W = 1, Z = 2.226$, $p < 0.05$, $r = 0.3934$), but in the opposite direction. Thus, less people self-assessed their stage of change in Contemplation. As we did not observe a significant change in Action stage nor in the overall intentions, we concluded that the interventions helped WEIZ's employees to be aware of the energy concerns, but no more.

4.2.4. ALLIA

Similar observations occurred in Cambridge, UK (ALLIA). The medians of PRE and POST regarding Pre-contemplation stage were -1.25 and -2.75, respectively. A Wilcoxon Signed-rank test shows that there was a significant effect as well ($W = 1$, $Z = 2.360$, $p < 0.05$, $r = 0.2905$). This means that ALLIA's employees equally changed their Pre-contemplation stage towards higher stages of consciousness.

4.2.5. MPH

According to the questionnaires, employees in Pilea-Hortiatis (Greece) reported statistical evidence that the mean of the Contemplation stage was different between PRE ($\mu = 1.074$) and POST ($\mu = -0.55$) ($W = 1, Z = 2.1388$, $p < 0.05$, $r = 0.356$). However, as not conclusive data for predecessor or subsequent stages were found, we can not conclude that MPH's employees improved or worsen their willingness to do actions in favour of energy efficiency.

A contingency analysis for the barriers using Fisher's test showed that significantly ($p < 0.01$) none of MPH's participants reported that their hurdles to behave energy efficiently were due to a lack of knowledge after the delivery of the treatments in POST (the evaluated item: "I am not sure about what is a good energy practice so I do little or nothing") nor people were found to be discouraged about peers at work (the evaluated item: "I am discouraged by the attitude of my colleagues and/or of the management, so I do little or nothing"). The following Table 5 provides an overview of these findings. These results on barriers were only reported in Pilea's pilot and not observed in other premises.

Table 5. Contingency table of MPH: PRE vs. POST in overall barriers to behave energy efficiently.

	Absentmindedness	Lack of Awareness	Peers' Discouragement	Other
PRE	6	5	5	10
POST	10	0	0	3

4.3. *Prescriptive Analysis*

The prescriptive analysis of the survey results aims at identifying and quantifying interactions between socio-economic data and persuasion strategies or other important behavioural constructs such as Pre-Contemplation, Contemplation, Action, Intentions, Barriers and Confidence in technology. To this aim, it was decided to use contingency tables which are popular in surveys' evaluation [32], also known as cross tabulation, as a means to understand whether the top rankings provided to PPs could be dependent on the variability of each socio-economic factor under study. Next, we provide these associations at two different snapshots separated by one year and a half: PRE and POST.

4.3.1. Results from Pre-Pilot Questionnaires

We evaluated if any of the fifteen Persuasive Principles were dependent on studied socio-economic variables (e.g., Gender: male/female). A Pearson's chi-squared test was applied to evaluate how likely it is that any observed difference between the variables within each factor arose by chance. To be even more rigorous and conservative, as multiple hypotheses were tested, a Bonferroni correction was applied on the significance levels for validating the hypotheses.

The results obtained can be observed in Table 6. The table shows that the only factors which show dependencies on some principles of persuasion were: City, Education, and Initiative_to_join. This latter factor, as well as City, were the ones which presented more significant interactions with different Principles of Persuasion. Specifically: Cause & effect (P2), Praise (P7), Similarity (P12) and Suggestion (P15).

Furthermore, we investigated if there were dependencies among the city of origin and certain important factors, such as Pre-Contemplation, Contemplation, Action, overall Intentions, Barriers and Confidence in technology. This test is relevant since it helps to observe if different employees respond differently to some factors under study depending on their country of origin, the city where they live or the working area where they are employed.

Table 6. Significant p-values (and the associated power) that ascertain the dependencies between factors and persuasive principles for the PRE survey.

Persuasion Principles	City	Education	Initiative_to_join
P2	$p < 0.0005$ $\beta = 0.9944237$	-	$p < 0.0005$ $\beta = 0.9339491$
P7	-	$p < 0.005$ $\beta = 0.9188609$	$p < 0.002$ $\beta = 0.9659801$
P12	$p < 0.001$ $\beta = 0.9764619$	-	-
P15	$p < 0.0005$ $\beta = 0.9996468$	-	-

Pre-Contemplation, Contemplation, and Action

A Kruskal–Wallis test between the city and these three factors revealed a significant effect of $(\chi^2(1) = 462.69, p < 0.01)$, $(\chi^2(1) = 285.7, p < 0.01)$ and $(\chi^2(1) = 303.3, p < 0.01)$. In Table 7 the results from a post hoc analysis is provided using Dunn's tests with Bonferroni correction.

Table 7. PRE-pilot: Post hoc analysis to identify where the differences are in the effect that we observed after running Kruskal-Wallis test. In Pre-contemplation it seems that Weiz scored differently to other pilots. Specifically, people in Weiz pilot scored higher the Pre-contemplation stage than the rest of the pilots.

	Pre-Contemplation	Contemplation	Action
City	Weiz-Bilbao Weiz-Seville Weiz-Thessaloniki Weiz-Cambridge Weiz-Sussex $(p < 0.002; r = 0.0875)$	Bilbao - Sussex $(p < 0.025; r = 0.058)$	-

Intentions, Barriers and Confidence in Technology

A contingency analysis using G test (The G test statistic is also approximately chi-squared distributed, but for small samples. This approximation is closer than one that chi-squared test uses) showed that Confidence in Technology $(\chi^2(14) = 25.763, p < 0.05)$ and the different barriers to behave pro-environmentally $(\chi^2(21) = 54.117, p < 0.01)$ also depend on the city significantly. In the following figures the reader can observe the distribution of different responses depending of the geographical distribution of pilots. Those images help identifying in a glimpse which of them responded differently. The area of tiles in the figures is proportional to the number of observations within each category (The reader can observe that number four is missing in the Figures (i.e., Figures 6 and 7). This is because we provided the codification of number four for other cities. However, as we did not get any observation from other locations, we show no data for number four).

According to the Pearson standardised residuals measures [32] performed over the contingency data in Figure 6, it indicates that the category Confidence in technology is negatively associated with people in Sussex pilot and No-Confidence positively associated with the employees of this UK building. Moreover, Seville's employees presents a positive association with No-Confidence and Thessaloniki's respondents presents a negative association with No-Confidence. These results means that in Sussex there are few people associated with Confidence in technology and many that are associated with No-confidence. The majority of Seville respondents are associated to No-Confidence, in contrast with Thessaloniki where just a few are associated to No-confidence (Pearson standardised residuals measure how large is the deviation from each cell to the null hypothesis (in this case, independence between row and column's). Please note that results with absolute value greater than two are significant indicative of association).

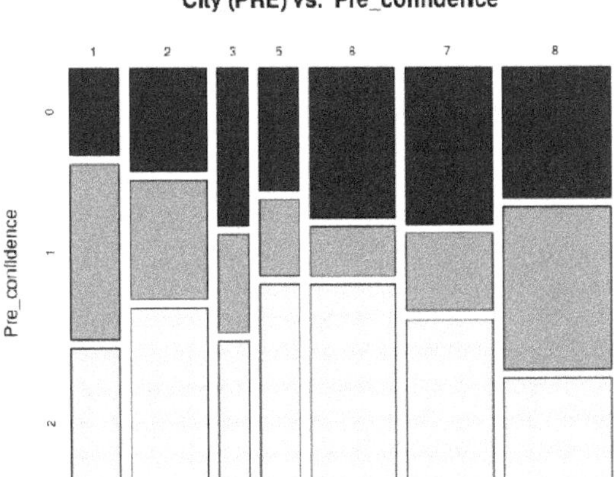

Figure 6. PRE-pilot: Comparison of the distribution of responses to confidence in technology among different pilots: neutral (0), No-Confidence (1) and Confidence (2).

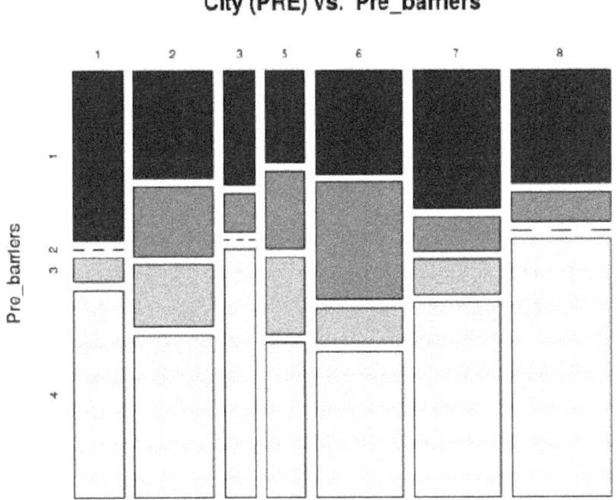

Figure 7. PRE-pilot: Comparison of the distribution to barriers encountered in each pilot: Absentmindedness (1), Lack of Awareness (2), Discouragement from Peers (3) and Other Reasons (4). In a glimpse we can observe that there are differences among certain pilots.

According to the Pearson standardised residuals measures performed over the contingency data in Figure 7, we observed a significant negative association of factor Lack of awareness in Seville pilot and a strong positive association of this factor in Thessaloniki. Positive association of Discouragement was found in Bilbao and MPH while a negative one was observed in Sussex. Finally, negative association of Other barriers was reported in Thessaloniki and Sussex.

4.3.2. Results from Post-Pilot Questionnaires

At the end of the intervention, a second assessment was performed to examine if the rankings provided to Persuasive Principles were still dependent on the variability of the socio-economic factors under study. Given the attrition rate on the POST survey we did not get any significant result on the Pearson's chi-squared test assuming the very restrictive Bonferroni correction. Removing this correction threshold, Table 8 reports the significant p-values found and the associated power for this particular analysis. In this case, Similarity (P12), Conditioning (P3), Social Recognition (P14) and Reciprocity (P9) were found to be dependent on three socio-economic variables. As can be observed, City appeared in both PRE and POST analyses as a dependent factor for certain Persuasive Principles. Specifically, we found a double repetition of this factor affecting Similarity (P12) principle of persuasion in both snapshots. Besides, Confidence in Technology appears to be the variable with more PP associations. In fact, it is related to Conditioning (P3), which was one of the Persuasion Principles used in the treatments delivered. This result was not observed in the PRE analysis (see Table 6), yet it is related to a previous finding of the authors of this article in a previous research[33]. That is, interventions that try to affect the behaviour on energy efficiency through ICT-based equipment has an impact on the Confidence in Technology of the people subjected to this kind of interventions.

Table 8. Significant p-values (and the associated power) that ascertain the dependencies between Factors and Persuasive Principles for the POST survey.

Persuasion Principles	City	Organisation Strategy	Confidence
P3	-	-	$p < 0.001$ $\beta = 0.6829225$
P9	-	-	$p < 0.01$ $\beta = 0.5013527$
P12	$p < 0.02$ $\beta = 0.9154948$	-	$p < 0.05$ $\beta = 0.6605258$
P14	-	$p < 0.05$ $\beta = 0.6637232$	-

As was done in the PRE-pilot, we have also investigated if there were dependencies among the different pilots (i.e., the cities where these are) and behavioural factors.

Pre-Contemplation, Contemplation and Action

As we observed in PRE, we continued finding differences among cities in these factors. A Kruskal–Wallis revealed a significant effect of City on Pre-Contemplation ($\chi^2(1) = 141.81, p < 0.01$), Contemplation ($\chi^2(1) = 58.559, p < 0.01$) and Action ($\chi^2(1) = 46.227, p < 0.01$). However, when we provided a post hoc test using Dunn's tests with Bonferroni correction we only saw the following significant differences reported in Table 9.

Table 9. POST-pilot: Post hoc analysis to identify where the differences are in the effect that we observed after running Kruskal Wallis test. In Pre-contemplation and Contemplation it seems that Bilbao scored differently than Greek pilots. Specificly, people in Greece scored higher on Pre-contemplation stage than the Spanish pilots. Interestingly, DEUSTO, SEVILLE and ALLIA reported the lowest scores in this stage of no-awareness. Regarding the Contemplation stage, there is a difference between the Spanish cities where DEUSTO again reported the highest average score in this stage of change and SEVILLE the lowest along with ALLIA and CERTH.

	Pre-Contemplation	Contemplation	Action
City	Thessaloniki-Bilbao Pilea-Bilbao ($p < 0.008; r = 0.254$)	Seville - Bilbao Thessaloniki-Bilbao Cambridge -Bilbao ($p < 0.01; r = 0.261$)	-

Intentions, Barriers and Confidence in Technology

A contingency analysis for these factors using G test, again because of the sample size, showed that all of them depend on the city significantly (whereas in PRE, Intentions did not depend on City).

Intentions ($\chi^2(12) = 23.735, p < 0.05$), Confidence ($\chi^2(12) = 38.48, p < 0.005$) and Barriers ($\chi^2(18) = 40.35, p < 0.002$). Figures 8–10 show the different distribution of responses depending of the city of origin that help us identifying which of them responded differently to the rest.

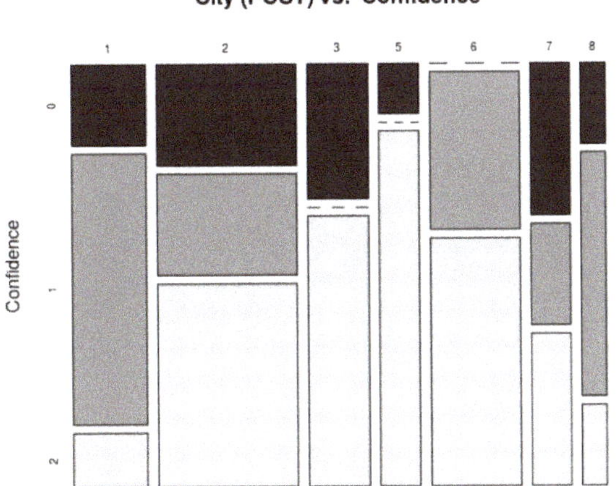

Figure 8. POST-pilot: Comparison of the distribution of confidence in technology among different pilots. Y axis: neutral (0), No Confidence (1) and Confidence (2).

According to the Pearson standardised residuals measures performed over the contingency data in Figure 8, we observed a negative association of Neutrality with Thessaloniki's employees. A strong positive association of No-confidence in technology reported in Seville pilot and a slight negative association in WEIZ building. Finally, we observed a strong negative association of the factor Confidence in technology in Seville while in MPH employees provided a positive association of this factor (i.e., MPH reported the highest rate of confidence in technology in a significant way).

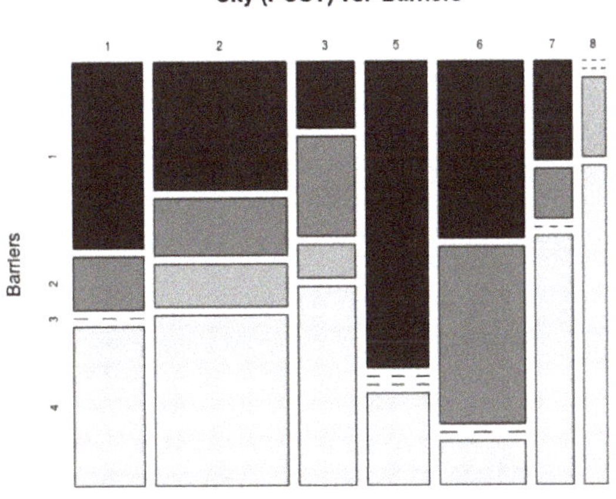

Figure 9. POST-pilot: Comparison of the distribution of barriers in each pilot. Y axis: Absentmindedness (1), Lack of Awareness (2), Discouragement from Peers (3) and Another Reason (4).

Pearson standardised residuals measures performed over the contingency data in Figure 9 provided a quite strong positive association of factor Absentmindedness in MPH and a strong positive association of factor Lack of awareness in Thessaloniki's employees. In this latter pilot, we also observed a negative association with Other barriers, meaning that tenants from this building identified clearly the barriers that hinder sustainable actions.

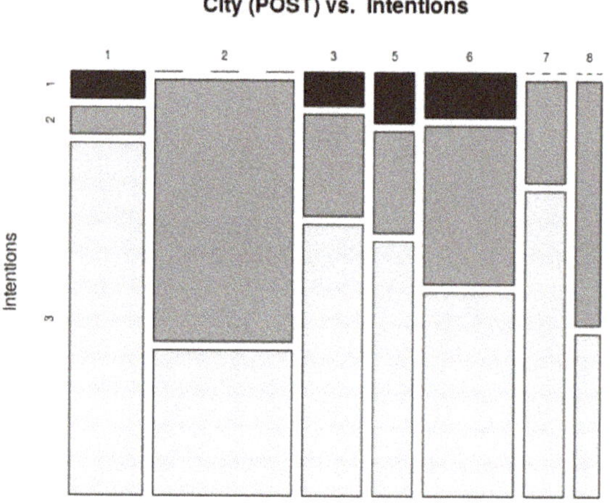

Figure 10. POST-pilot: Comparison of the distribution of intentions in each pilot. Y axis: Pre-Contemplation (1), Contemplation (2) and Action (3).

To conclude this analysis, Pearson standardised residuals measures performed over the contingency data in Figure 10 reported a positive association of Contemplation and a slight negative association of Action in Bilbao's building (DEUSTO). In Seville, however, we found a positive association of Action which means that the majority employees in this building reported to do actions in favour to the environment.

4.4. Summary of Results

Both the PRE and POST pilot surveys have been analysed in a descriptive and prescriptive manner. By comparing the results of the two surveys, the overall characteristics of the population remained more or less the same as has been reported in Section 4.2. It was really an interesting finding to observe that the results from the POST questionnaire pointed out to the same top three strategies that were best ranked in PRE: (Cause and Effect (P2), Conditioning (P3) and Self-monitoring (P11)). Indeed, these three strategies were the ones selected after the first intervention to inform the design of the different treatments delivered during the GreenSoul study.

Regarding the socio-economic factors that affect self-perceived usefulness of Persuasive Principles, "City", "Education" and "Initiative to join environmental campaigns" were identified as dependent factors over some Persuasion Principles in Pre-Pilot questionnaire. In the POST questionnaire, only City remained as an affecting factor to some PPs. Besides, we found that Confidence in technology and Organisational factors at workplace appeared to provide an effect in POST that were not observed in PRE. The Similarity principle ("People are more readily persuaded through systems that remind them of themselves in some meaningful way.") was the only PP observed in the results of both questionnaires always affected by City.

In terms of intentions to act in favour of the environment, it was relevant to find that general employees' pro-environmental intentions have shifted towards more active participation (Action) than before the pilot intervention by 8.5%. Statistically speaking, we only found a difference in the Pre-contemplation stage between the responses of PRE and POST pilots questionnaires when treating the dataset as a whole (pooled samples). Thus, in general, at the end of the ICT-based intervention employees seemed to be more aware of the potential actions to do towards energy efficiency.

Furthermore, we performed an analysis of all the behavioural constructs break down by pilot (i.e., comparing previous and post responses to the questionnaire within different cities). On the one hand, such analysis helped improve our understanding of where the reported statistical difference in Pre-Contemplation was more accentuated. On the other hand, we were able to observe other differences in behavioural constructs that were not noticeable when doing pooled analysis. In DEUSTO (Bilbao, Spain), we found an strong positive difference in the Pre-contemplation stage between the responses provided in PRE and POST. Furthermore, employees in this pilot increased their scores in Contemplation and Action stages. This means that the overall effect of the ICT-based intervention was effective towards the purpose of making people to be more aware about the environment and contributing to mitigate energy inefficiency in an active manner. In SEVILLE (Spain), people increased their overall intentions in favour of the environment and it was the pilot with the higher number of employees responding to be in Action stage. WEIZ (Austria) and ALLIA (Cambridge, UK) also changed their Pre-contemplation stage towards higher stages of consciousness. In MPH (Pilea-Hortiatis, Greece), we did not find conclusive information on employees' intentions to behave in a more energy efficient manner. In the results, we do not provide information about ECOLUTION (Sussex, UK) nor CERTH (Thessaloniki, Greece) because of we did not find any differences on their results reported in PRE and POST time-frames. This result is relevant as these pilot sites were control conditions. Therefore, the potential effect of GreenSoul's intervention described in Section 2.1 is reinforced as no differences between PRE and POST were found on those pilots-buildings.

Barriers and Confidence in technology have been thoroughly studied in the analysis provided. At the end of the intervention, the majority of the interviewees reported to have less barriers to act pro-environmentally than in PRE stage. Besides, having a look at the contingency tables in PRE and POST we found that MPH moved from discouragement to absentmindedness as a primary factor hindering the adoption of energy efficient actions. In this Greek building, none of the respondents reported that one of the obstacles to behave energy efficiently were due to lack of knowledge nor people were found to be discouraged about their work-peers. This finding suggests that the intervention might have increased the energy-awareness significantly and reduced the discouragement about peers. If the main barrier, overall, that users encounter after the end of the intervention in MPH was occasional absentmindedness and not other, our hypothesis is that the GS intervention should have provided a positive effect. In the analysis we also observed how CERTH (control group) provided a strong association with the factor Lack of awareness in PRE and POST. This finding is indicative that these barriers steadily prevent employees to behave pro-environmentally at the workplace if they are not provided with any green-intervention.

Finally, we reported the impact of the Confidence in technology in the success or failure of ICT-based persuasive interventions. In POST, we found that "trust" seems to be an affecting factor for three PPs: Conditioning, Reciprocity and Similarity. As this finding was not observed in PRE questionnaire, we hypothesise that the GS intervention mediated this effect. However, this is only a exploratory result that deserves further investigation. In the same vein, it was also interesting to observe that MPH was the Pilot that increased its confidence in technology by far with regards to other pilots after the intervention. Conversely, SEVILLE's respondents always presented less confidence in technology at the beginning and end of the piloting regardless of the ICT-based intervention.

5. Implications on ICT-Based Interventions for Sustainability

The study has provided relevant results in terms of the design of future interventions based on ICT towards energy-efficiency in tertiary buildings. On the one hand, the results pointed out that Cause and Effect (P2), Conditioning (P3) and Self-monitoring (P11) were the top self-ranked Persuasive

Principles across all pilot buildings in UK, Spain, Greece and Austria. Observing that even after the intervention the employees kept choosing them as the top-strategies, that suggested that we were right selecting and delivering them in the different ICT-based experimental conditions. In the body of the literature, other researchers have reported that other principles of persuasion were equally effective for different user-profiles (e.g., praise [34], commitment [35], social approval [36] or normative feedback [37]). This confirmed that different socio-economic factors affect the self-reported principles of persuasion. Therefore, it is challenging to provide gold-standards without performing ex-ante user-research studies of the population that will receive them [38].

According to the results reported in this manuscript, we argue that the success or failure of an ICT-based strategy based on Similarity principle will depend on the city where it will be delivered. This finding is controversial as "City" is just a physical location. Therefore, we deem that behind the City factor, there are many hidden cultural factors that have not been captured in this questionnaire but that are affecting principles of persuasion (in this case, Similarity). In any case, the result appears to be right as sharing similar attitudes and traits or sharing membership in a group was found to be positively associated with liking the persuader and endorsing the persuasive message [39]. In fact, a study found that demographic and behavioural similarity between the source and recipients resulted in more positive behavioural changes [40].

The discouragement about peers as a barrier for doing pro-environmental actions was reduced after the intervention. This finding helps to demystify the idea that external factors play one of the key roles on demotivating people at workplace [9]. Besides, some pilots have completely changed the barriers encountered to behave energy-efficiently, which did not occur in control groups (e.g., CERTH). Interestingly in the same country, all MPH's employees reported a full understanding of the actions to do in favour of the environment. However, they regularly present absentmindedness. Therefore, it seems that frequent subtle feedback should be provided to employees and tenants to reduce forgetting once they are aware of the problem.

Finally, the results suggest that it is important to know the level of confidence in technology if an ICT-based intervention to change the people's behaviour was based on Conditioning, Reciprocity and Similarity principles. Thus, their success or failure might partially depend on the level of confidence in the technology of the end-users. Faith or trust on technology is usually well perceived by users almost without objection [41]. However, it is interesting to note some technological paradoxes in this context. Technology has a leading role as a solution but sometimes it is also part of the problem (e.g., technological artefacts conceived towards energy efficiency that do not really compensate during their lives the greenhouse gases emitted to produce them). Moreover, enhancements on energy efficiency can provoke increased demand for energy services or even misuse of them (Jevons paradox [42]). In the same vein, over-reliance on cutting-edge technology may bring undesired effects to pro-environmental behaviour and reduce the personal responsibility for action [43]. Conversely, we did not found these reported effects from the body of literature. In our study the pilots with higher level of confidence in technology at the end of the intervention (specially MPH) were found to be the ones with less barriers to behave energy efficiently or which intentions to do actions in favour of the environment were enhanced.

Limitations and Mitigation Actions

In this study we have provided useful insights about how user profiles can be used to inform more efficient ICT-based campaigns to promote sustainable practices in tertiary buildings. However, the work has some limitations that should be taken into consideration for generalisation purposes. First, the reduction observed from the PRE to the POST-pilot responses (one third) was high. However, it reflects typical attrition rates in similar research. In similar studies that involve ICT solutions, their usage is mostly at the discretion of the participant, and the participant has the option to discontinue its usage very easily. In any longitudinal study, where the intervention is neither mandatory nor critical to the participants daily activities (well-being, pro-environmental sustainability or energy efficiency), trial participants will be lost well below of the 60–70% [44]. Although these rates were expected at the beginning of the study, we can not overlook that this represents the main limitation of the results. Beside the sample size reduction, the geographical distribution of the sample between PRE and POST (see Table 4) can be considered as another limitation. This latter issue might an effect on the reliability of the results presented; overall, the conclusions raised with pooled data (i.e., differences in Pre-Contemplation between PRE and POST). As a mitigation argument, we have declared and demonstrated with data that the overall sample was kept balanced in most socio-economic aspects (age, sex, education, etc.) between the two time-frames. More importantly, we provided an in-depth study within each pilot site in PRE and POST to draw specific conclusions for each pilot which reinforce the validity of the study. We acknowledge that with the data provided, it is not very easy to extrapolate results to the whole pan-European context. However, we argue that our results can shed light or give hints to designers, engineers, managers or other relevant stakeholders of tertiary buildings to decide which kind of interventions based on ICT and persuasion could be delivered according to their specific context and socio-economic profile. For that, in Appendix A, we provide a description of each building and the employees that work within.

Besides the previous shortcomings, hereafter we provide other limitations of the study which are typical on interventions that aim at forming new behaviours in the field. Lasting effect was not sufficiently addressed in the presented work. We have reported that subjects seem to have modified their behaviours at the end of the study. However, we were not able to measure whether the changes remained after the second snapshot. Future studies should address this limitation by providing washout periods at the beginning and at the end of the interventions. The selection of the population also presented difficulties to generalise results. The subjects participating in this study were selected using a RCT approach, and they were restricted to the participating entities in the European project. This approach could result in a bias for the generalisation of results since they might not be comparable to other cases in Western societies. Nevertheless, we found interesting results in public buildings where civil servants work (e.g., city councils). Therefore, we argue that the results may have an impact on this specific tertiary buildings in which schedules are fixed and there are groups of employees split in several offices were visitors get in and out daily. Finally, we report on the non-controlled effects that may hinder the internal validity of the study. It is difficult to be validated as we were investigating in a non-controlled environment. In-the-wild research related to sustainability usually entails several hidden and uncontrollable factors (e.g., global climate campaigns, pro-environmental media or news on climate emergency) which are out of the control of the researchers and may have had an impact on the responses the employees gave to the questionnaires. In essence, it is usual that researchers working on field studies report similar limitations as we have brought about in this section. It is important to emphasise the limitations, but also the validity and value that the empirical evidence provided in this manuscript have to the research community working on sustainability and pro-environmental behaviour change.

6. Conclusions and Outlook

To deliver a socio-economic tool that can be used to assess persuasion interventions in tertiary buildings, a two-step survey was designed and conducted in multiple buildings across Europe. Between the two steps, different persuasion strategies have been deployed in different premises through the GreenSoul project, which allowed the evaluation and validation of the survey as an assessment instrument. The results shed light on the importance of understanding user profiles both in socio-economic and behavioural terms to inform ICT-based campaigns to promote sustainable practices among employees. Beyond monetary incentives, which usually work in households, other engaging mechanisms need to be considered at tertiary buildings where employees are not aware of the impact of their everyday actions. Thus, recognition, certification, similarity or other carefully designed nudges to promote behavioural change. This work has found that a representative sample from the pan-European landscape agrees on selecting Self-monitoring, Cause and effect and Conditioning as the most promising principles to engage people into energy-efficiency. As there are already quite a few studies exploring which persuasion approaches seem to work best on different contexts, our future steps will seek to understand if people prefer a variety of incentivisation mechanisms or if they stick with just a few of them. Besides, we will emphasise on understanding how the confidence in technology, as a mechanism to solve all environmental issues, impact on the type of engaging mechanisms to deliver.

Author Contributions: Conceptualisation, D.C.-M., A.C.T. and C.E.B.; methodology, D.C.-M. and O.K.-E.; validation and formal analysis, D.C.-M. and A.C.T.; writing—original draft preparation, D.C.-M. and A.C.T.; writing—review and editing, S.K., D.L.-d.-I. and J.M.A.; supervision, D.T. and D.L.-d.-I.; project administration, J.M.A. All authors have read and agree to the published version of the manuscript.

Funding: This work has been partially supported by the European Commission through the project HORIZON 2020-RESEARCH & INNOVATION ACTIONS (RIA)-696129-GREENSOUL. We also acknowledge the support of the Spanish government for SentientThings under Grant No. TIN2017-90042-R.

Acknowledgments: The authors would like to thanks all those that helped in the pilot deployments and follow ups. Special mention to Arazt Manterola and Mikel Solabarrieta.

Conflicts of Interest: The authors declare no conflict of interest.

Abbreviations

The following abbreviations are used in this manuscript:

ICT	Information and Commu Technologies
GS	GreenSoul
PP	Persuasive Principle
PS	Persuasive Strategy
TTM	Transtheoretical Model
PRE	Pre-Pilot
POST	Post-Pilot
RCT	Randomised Controlled Trial

Appendix A. Offices and Pilots Description

Throughout the whole manuscript we provide relevant outcomes from official GreenSoul pilot buildings. Therefore, this section shows a descriptive summary of the pilot sites aside with relevant data that help to extrapolate results to similar settings (a thorough description of these buildings and characteristics from employees and tenants can be found in [45]).

WEIZ (Austria): The pilot was deployed at the Energy and Innovation Centre of Weiz (W.E.I.Z.), which is an innovative and trendsetting business centre. Here, 34 entrepreneurs and organisations in the field of F&E, Economy and Education find attractive office and storage rooms, which are conceived after the latest cognitions and get professional support from the management of W.E.I.Z. The W.E.I.Z. is located in the centre of Weiz (Austria). The optimal infrastructure is supplemented by flexible room sizes and a sophisticated use and energy concept. The building campus is the largest Styrian impulse centre outside the capital city Graz. The opening hours for visitors are from 07:00 a.m. to 07:00 p.m. on weekdays and 08:00 a.m. to 12:00 a.m. on Saturdays (in case of a specific event opening hours are extended to 11:00 p.m.). The tenants have keys and can access at any time thus their occupancy can only be estimated.

ALLIA (UK): The Future Business Centre is owned and operated by Allia Ltd. It opened in November 2013. The Future Business Centre is a business innovation centre with a difference—to grow businesses that do good for society and the environment. More than just a set of workspaces for rent, it is a place where people can grow their ideas to make a difference in the world. It offers affordable, high quality workspace on flexible terms with specialist business support and an ethos of collaboration and innovation. The building has shared toilets, changing facilities and tea points. A reheat kitchen and eating area are located on the ground floor. There are currently 500 live "access cards" but these include contractors "hot deskers" and a sizeable "part-time staff".

ECOLUTION (UK): Affinity Sutton is one of the largest providers of affordable housing in England managing over 58,000 homes and properties in over 120 local authorities. They are a non-profit organisation for social purpose with commitments to reduce carbon emissions and increase energy efficiency across its own buildings and housing stock including tackling fuel poverty of residents. The pilot is settled in an open space area on the third floor of the Upton House, a three-storey-building, comprising offices and meeting rooms. Specifically, the pilot area is an non automated open space with several groups of workstations distributed across the floor. It is estimated that employees spend ~6 h in the office, 1 h in meetings and 1 h at lunch. Usually there are approximately 10 visitors per day.

DEUSTO (Spain): The pilot was held at ESIDE building, which was built in 1921 and houses the famous Faculty of Economics and Business Administration. Its neoclassical façade is 107 m in length, and it consists of a basement, ground floor and two floors. In 1996, the modern building was attached to its back. ESIDE hosts the Faculty of Engineering, DeustoTech (a research centre which belongs to Deusto Foundation) and DeustoKabi (a start-up incubator). Moreover, the new adjacent facilities for the Sports Degree have been opened in 2014. All these buildings share the electricity, water and heating systems. The people in the pilot range from researchers, technicians, project managers, accountants, working an average of 8 h per day. The daily operation hours of the building are weekdays from 08:00 a.m. to 08:00 p.m. and on the weekends from 09:00 a.m. to 02:00 p.m. The employees can access the building at this time. The daily occupancy is estimated on 120 employees and 10 visitors per day.

SEVILLE (Spain): The pilot was held at the Institute of Statistics and Cartography of Andalusia, which was built in 1992 for the World Exposition that was hosted in Seville. The building was the New Zealand Pavilion that was built as a touristic building. After the end of the World Exposition, the pavilion was acquired by the public regional government and they built offices inside to host the whole Institute of Statistics and Cartography of Andalusia. The building consists of basement, three floors and roof. The pilot has an area of 3.529 m^2. One of the main challenges was to rebuild the building into offices and the major project was related to the adaptation of the air conditioning to current usage conditions and energy requirements. Due to the fact that the existing facility was designed for an unlimited period of time and characteristics of use, it was necessary to design a completely new air conditioning system. The offices are used by the public administrators and civil servants. The vast majority of this personnel spend at least 7 h in the office.

MPH (Greece): The pilot was held at the Pilea-Hortiatis municipality hall buildings. The Municipality of Pilea-Hortiatis is based in a quite new building operating from 2010 on. A variety of retrofitting actions have been performed, updating the buildings to energy category B according to KENAK (Greek Regulation for the Energy Efficiency of Building), reducing their operational energy consumption. Furthermore, the roof of the Municipality Hall buildings is partially covered with photovoltaic (PVs) panels and the energy produced is sold to the National Electricity Provider Company (Public Power Corporation S.A). The Municipal Hall is open to the general public 5 days a week from 7:00 a.m. to 16:00 p.m., while Building B is open in the afternoons till 21:00 a.m. as it hosts the music school and the concert and conference halls. The offices are used by the administrative personnel and members of the Municipal Council or civil servants.

CERTH (Greece): The pilot was held at the Information Technologies Institute (ITI) central building at the Centre for Research Technology Hellas in Thessaloniki. The building was constructed in 2000, and consists of ground floor, two floors (one of which was constructed later as an extension with a metal foundation) and two undeground parking levels. Average working hours are from 09:00 to 17:00 and hosts mainly ICT-related activities, including also administrative ones. The building has also server rooms, dedicated labs, and meeting/conference rooms that are used based on daily needs. Even though there is a BMS available, all building assets (HVAC, Lights, Appliances) are fully controlled by the end-users.

References

1. Scherbaum, A.; Popovich, P.; Finlinson, S. Exploring individual-level factors related to employee energy-conservation behaviors at work. *J. Appl. Soc. Psychol.* **2008**, *38*, 818–835. [CrossRef]
2. Burger, P.; Bezençon, V.; Bornemann, B.; Brosch, T.; Carabias-Hütter, V.; Farsi, M.; Hille, S.; Moser, C.; Ramseier, C.; Samuel, R.; et al. Advances in understanding energy consumption behavior and the governance of its change–outline of an integrated framework. *Front. Energy Res.* **2015**, *3*, 29. [CrossRef]
3. Ghazali, E.; Nguyen, B.; Mutum, D.; Yap, S. Pro-environmental behaviours and Value-Belief-Norm theory: Assessing unobserved heterogeneity of two ethnic groups. *Sustainability* **2019**, *11*, 3237. [CrossRef]
4. Zhang, Y.; Bai, X.; Mills, F.P.; Pezzey, J.C. Rethinking the role of occupant behavior in building energy performance: A review. *Energy Build.* **2018**, *172*, 279–294. [CrossRef]
5. Paone, A.; Bacher, J.P. The impact of building occupant behavior on energy efficiency and methods to influence it A review of the state of the art. *Energies* **2018**, *11*, 953. [CrossRef]
6. Ajzen, I. The theory of planned behavior. *Organ. Behav. Hum. Decis. Process.* **1991**, *50*, 179–211. [CrossRef]
7. Hines, J.; Hungerford, H.; Tomera, A. Analysis and synthesis of research on responsible environmental behavior: A meta-analysis. *J. Environ. Educ.* **1987**, *18*, 1–8. [CrossRef]
8. Stren, P. Toward a coherent theory of environmentally significant behaviour. *J. Soc. Issues* **2000**, *56*, 407–424. [CrossRef]
9. Kollmuss, A.; Agyeman, J. Mind the gap: Why do people act environmentally and what are the barriers to pro-environmental behavior? *Environ. Educ. Res.* **2002**, *8*, 239–260. [CrossRef]
10. Belafi, Z.; Hong, T.; Reith, A. A critical review on questionnaire surveys in the field of energy-related occupant behaviour. *Energy Effic.* **2018**, *11*, 2157–2177. [CrossRef]
11. Thaler, R.; Sunstein, C. Nudge: Improving Decisions About Health, Wealth, and Happiness; Penguin. 2009. Available online: https://www.penguin.co.uk/books/567/56784/nudge/9780141040011.html (accessed on 12 November 2019).
12. Fogg, B.J. Persuasive technology: Using computers to change what we think and do. *Ubiquity* **2002**, *2002*, 2. [CrossRef]
13. Cialdini, R.B.; Cialdini, R.B. *Influence: The Psychology of Persuasion*; Morrow: New York, NY, USA, 1993.
14. Froehlich, J.; Findlater, L.; Landay, J. The design of eco-feedback technology. In Proceedings of the SIGCHI Conference on Human Factors in Computing Systems, Atlanta, GA, USA, 10–15 April 2010; pp. 1999–2008. [CrossRef]
15. Kosters, M.; Van der Heijden, J. From mechanism to virtue: Evaluating Nudge theory. *Evaluation* **2015**, *21*, 276–291. [CrossRef]
16. Hermsen, S.; Frost, J.; Renes, R.J.; Kerkhof, P. Using feedback through digital technology to disrupt and change habitual behavior: A critical review of current literature. *Comput. Hum. Behav.* **2016**, *57*, 61–74. [CrossRef]
17. De Young, R. Changing behavior and making it stick: The conceptualization and management of conservation behavior. *Environ. Behav.* **1993**, *25*, 485–505. [CrossRef]
18. Knowles, B.; Blair, L.; Walker, S.; Coulton, P.; Thomas, L.; Mullagh, L. Patterns of persuasion for sustainability. In Proceedings of the 2014 Conference on Designing Interactive Systems (DIS'14), Vancouver, BC, Canada, 21–25 June 2014; pp. 1035–1044. [CrossRef]
19. GreenSoul, H2020. Available online: http://www.greensoul-h2020.eu/ (accessed on 2 April 2020).
20. Casado-Mansilla, D.; Moschos, I.; Kamara-Esteban, O.; Tsolakis, A.C.; Borges, C.E.; Krinidis, S.; Irizar-Arrieta, A.; Konstantinos, K.; Pijoan, A.; Tzovaras, D.; et al. A Human-Centric Context-Aware IoT Framework for Enhancing Energy Efficiency in Buildings of Public Use. *IEEE Access* **2018**, *6*, 31444–31456. [CrossRef]
21. Casado-Mansilla, D.; Lopez-de Armentia, J.; Garaizar, P.; López-de Ipiña, D. To switch off the coffee-maker or not: That is the question to be energy-efficient at work. In Proceedings of the CHI'14 Extended Abstracts on Human Factors in Computing Systems, Toronto, ON, Canada, 26 April–1 May 2014; pp. 2425–2430. [CrossRef]
22. Whittle, C.; Jones, C. User perceptions of energy consumption in university buildings: A University of Sheffield case study. *J. Sustain. Educ.* **2013**, *5*, 27.

23. Casado-Mansilla, D.; Papageorgiou, D.; Borges, C.; Tsolakis, A.; Kamara, O.; Krinidis, S.; Sanchez, R.; Moschos, I.; Zacharaki, A. (PRE) Socio-economic and cultural dataset in relation to Persuasive Strategies to boost Energy Efficiency and in the UK, Spain, Greece and Austria. 2019. Available online: http://10.5281/zenodo.2610102 (accessed on 23 February 2020).
24. Papageorgiou, D.; Casado-Mansilla, D.; Tsolakis, A.C.; Borges, C.E.; Zacharaki, A.; Kamara-Esteban, O.; Avila, J.M.; Moschos, I.; Irizar-Arrieta, A. A Socio-economic Survey for Understanding Self-perceived Effectiveness of Persuasive Strategies Towards Energy Efficiency in Tertiary Buildings. In Proceedings of the 5th IEEE International Conference on Internet of People, Leicester, UK, 19–23 August 2018; pp. 1–6.
25. Oinas-Kukkonen, H.; Harjumaa, M. Persuasive Systems Design: Key Issues, Process Model and System Features 1. In *Routledge Handbook of Policy Design*; Routledge: London, UK, 2018; pp. 87–105.
26. Casado-Mansilla, D.; Tsolakis, A.; Borges, C.; Manterola, A.; Sanchez, R.; Krinidis, S.; Kamara, O.; López de Ipiña, D. (POST) Socio-economic and cultural dataset in relation to Persuasive Strategies to boost Energy Efficiency and in the UK, Spain, Greece and Austria. 2019. Available online: http://10.5281/zenodo.3565757 (accessed on 23 February 2020).
27. Lockton, D.; Harrison, D.; Stanton, N.A. Models of the user: Designers' perspectives on influencing sustainable behaviour. *J. Des. Res.* **2012**, *10*, 7–27. [CrossRef]
28. Tribble, S.L. Promoting environmentally responsible behaviors using motivational interviewing techniques. In *Honors Projects*; Illinois Wesleyan University: Bloomington, IL, USA, 2008.
29. Sutton, S. Transtheoretical model of behaviour change. In *Cambridge Handbook of Psychology, Health and Medicine*; Cambridge University Press: Cambridge, UK, 2007; pp. 228–232. Available online: https://doi.org/10.1017/CBO9780511543579.050 (accessed on 23 February 2020).
30. Sullivan, G.M.; Feinn, R. Using effect size—or why the P value is not enough. *J. Grad. Med Educ.* **2012**, *4*, 279–282. [CrossRef]
31. Eichhorn, W. Fisher's tests revisited. *Econom. J. Econom. Soc.* **1976**, 247–256. [CrossRef]
32. Agresti, A. A survey of exact inference for contingency tables. *Stat. Sci.* **1992**, *7*, 131–153. [CrossRef]
33. Casado-Mansilla, D.; Garaizar, P.; López-de Ipiña, D. User Involvement Matters: The Side-Effects of Automated Smart Objects in Pro-environmental Behaviour. In Proceedings of the 9th International Conference on the Internet of Things, Bilbao, Spain, 22–25 October 2019; pp. 1–4. [CrossRef]
34. Anagnostopoulou, E.; Magoutas, B.; Bothos, E.; Schrammel, J.; Orji, R.; Mentzas, G. Exploring the links between persuasion, personality and mobility types in personalized mobility applications. In *International Conference on Persuasive Technology*; Springer: Berlin, Germany, 2017; pp. 107–118.
35. Oyibo, K.; Adaji, I.; Orji, R.; Olabenjo, B.; Vassileva, J. Susceptibility to persuasive strategies: A comparative analysis of Nigerians vs. Canadians. In Proceedings of the 26th Conference on User Modeling, Adaptation and Personalization, Singapore, 8–11 July 2018; pp. 229–238.
36. Oyibo, K.; Orji, R.; Vassileva, J. Investigation of the Persuasiveness of Social Influence in Persuasive Technology and the Effect of Age and Gender. In *PPT@ PERSUASIVE*; Amsterdam, The Netherlands. 2017; pp. 32–44. Available online: http://ceur-ws.org/Vol-1833/8_Oyibo.pdf (accessed on 3 April 2020).
37. Peschiera, G.; Taylor, J.E.; Siegel, J.A. Response–relapse patterns of building occupant electricity consumption following exposure to personal, contextualized and occupant peer network utilization data. *Energy Build.* **2010**, *42*, 1329–1336. [CrossRef]
38. Irizar-Arrieta, A.; Casado-Mansilla, D.; Garaizar, P.; López-de Ipiña, D.; Retegi, A. User perspectives in the Design of Interactive Everyday Objects for Sustainable Behaviour. *Int. J. Hum. Comput. Stud.* **2020**, *137*, 102393. [CrossRef]
39. Kim, M.; Shi, R.; Cappella, J.N. Effect of character–audience similarity on the perceived effectiveness of antismoking PSAs via engagement. *Health Commun.* **2016**, *31*, 1193–1204. [CrossRef] [PubMed]
40. Durantini, M.R.; Albarracin, D.; Mitchell, A.L.; Earl, A.N.; Gillette, J.C. Conceptualizing the influence of social agents of behavior change: A meta-analysis of the effectiveness of HIV-prevention interventionists for different groups. *Psychol. Bull.* **2006**, *132*, 212. [CrossRef] [PubMed]
41. Ponce, P.; Polasko, K.; Molina, A. End user perceptions toward smart grid technology: Acceptance, adoption, risks, and trust. *Renew. Sustain. Energy Rev.* **2016**, *60*, 587–598. [CrossRef]
42. Alcott, B. Jevons' paradox. *Ecol. Econ.* **2005**, *54*, 9–21. [CrossRef]

43. Murtagh, N.; Gatersleben, B.; Cowen, L.; Uzzell, D. Does perception of automation undermine pro-environmental behaviour? Findings from three everyday settings. *J. Environ. Psychol.* **2015**, *42*, 139–148. [CrossRef]
44. Eysenbach, G. The law of attrition. *J. Med. Internet Res.* **2005**, *7*, e11. [CrossRef]
45. GreenSoul Project. Greensoul Pilots. Available online: http://www.greensoul-h2020.eu/public-deliverables/greensoul-end-users-requirements-report (accessed on 23 March 2020).

Article

A National Strategy Proposal for Improved Cooking Stove Adoption in Honduras: Energy Consumption and Cost-Benefit Analysis

Wilfredo C. Flores [1], Benjamin Bustamante [2], Hugo N. Pino [1], Ameena Al-Sumaiti [3] and Sergio Rivera [4,*]

1 Faculty of Engineering, Universidad Tecnológica Centroamericana, UNITEC, 11101 Tegucigalpa, Honduras; wilfredo.flores@unitec.edu.hn (W.C.F.); hugo.pino@unitec.edu.hn (H.N.P.)
2 Associate Consultor, Ingenieria ByB Asociados, 11101 Tegucigalpa, Honduras; bbustamente@bybasociados.com
3 Advanced Power and Energy Center, Electrical Engineering and Computer Science Department, Khalifa University, P.O. Box 127788 Abu Dhabi, UAE; ameena.alsumaiti@ku.ac.ae
4 Electrical Engineering Department, Universidad Nacional de Colombia, 111321 Bogotá, Colombia
* Correspondence: srriverar@unal.edu.co

Received: 30 December 2019; Accepted: 12 February 2020; Published: 19 February 2020

Abstract: The high consumption of firewood in Honduras necessitates the search for alternatives with less-negative effects on health, the economy, and the environment. One of these alternatives has been the promotion of improved cooking stoves, which achieve a large reduction in firewood consumption. This paper presents a cost-benefit analysis for an improved cooking stove adoption strategy for Honduras. The methodology uses the *Long-range Energy Alternatives Planning System*, LEAP, a tool used globally in the analysis and formulation of energy policies and strategies. The energy model considers the demand for firewood as well as the gradual introduction of improved cooking stoves, according to the premises of a National Strategy for improved cooking stoves adoption in Honduras. Hence, it is demonstrated that the costs of implementing this adoption strategy are lower than the costs of not implementing it, taking into consideration representative scenarios up to and including the year 2030.

Keywords: cost-benefit analysis; energy strategy; improved cook stoves; Honduras

1. Introduction

Firewood is a very important source of energy in Honduras [1]. Many households with access to electricity still use firewood as the main source of energy for cooking food. Firewood is also used in micro and medium enterprises dedicated to the sale of food, salt extraction, brick production, bakeries, tortilla manufacturing, and coffee mills, among others. In urban and peri-urban areas, 29% of households use firewood, while in rural areas firewood continues to predominate in 88% of households [1]. Hence, in the last few decades there has been a significant increase in deforestation in Honduras. Studies reveal that the volume loss per year is 58,000 hectares. In 2015, after a period of 17 years, the forest reduction was 870,000 hectares [2].

Energy is essential for human development in various ways, such as health care, transportation, information, communication, lighting, heating, food processing, and other uses. Therefore, energy poverty has serious implications for basic human needs, such as cooking, heating the home, lighting or access to basic media services.

In the Honduran case, according to the use of energy in households, the total number of basic energy needs is six. Figure 1 shows these six groups of basic energy needs for Honduran households.

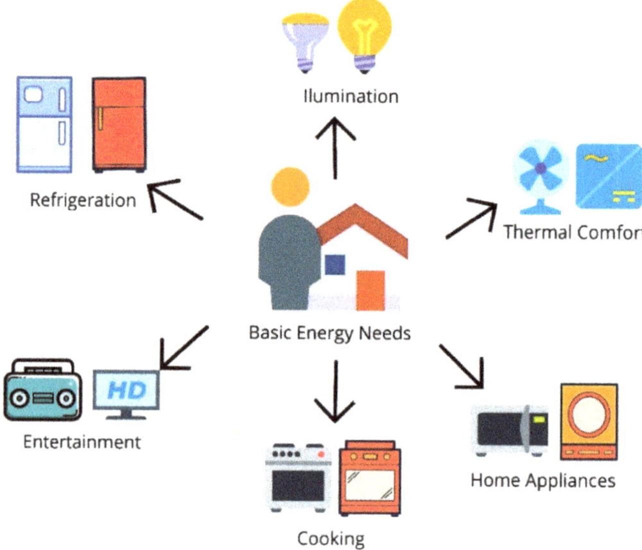

Figure 1. Basic energy needs for Honduran households.

The majority of poor countries around the world use firewood to meet some of these needs, mainly for cooking. In many cases, the use of biomass is not the most appropriate or suitable in terms of implications for health, and they are not precisely cheaper sources, but they tend to be the only option available. Despite the potential of technologies such as solar ovens [3] and others to be useful, a large quantity of developing countries still use firewood for cooking.

About half of Honduran households (approximately one million) cook with traditional wood-burning stoves [1,4]. These stoves are not only inefficient, but also have highly detrimental effects on the health of the user. In addition, the cost of collecting or buying firewood also has a huge impact on the economy and social welfare of families.

Consequently, the high consumption of firewood in Honduras requires the search for alternatives that reduce its negative impacts. In the country, one of these alternatives has been the promotion of improved stoves. This adoption achieves a large reduction in firewood consumption, as improved stoves can potentially use up to 71.2% less wood than traditional stoves, depending on the technology and user [4]. Additionally, families cooking with a traditional stove in zones where it is difficult to find firewood (peri-urban areas) spend about USD 20.00 per month on firewood purchases. Furthermore, also exists the cost of travel and time for its collection. Additionally, it is necessary to take the health expenses of respiratory diseases associated with the traditional stoves into account [5,6]. Figure 2 shows traditional (**a**) and improved stoves (**b**) used in Honduras.

However, the country programs that introduced improved stoves have traditionally been isolated efforts, with few resources for technological development and with a lack of follow-up on the adoption of new technologies [6]. Additionally, the adoption of improved stoves in Honduran households has been affected by a lack of public policies or strategies with a long-term vision for the development of a value chain that integrates the different links, such as design, manufacturing, financing, marketing, and post-sales services, as well as the sustainable supply of wood [7].

In this way, a change of direction is required; it calls for a comprehensive and joint strategy that allows the use of improved stoves to develop under different conditions. This strategy must be economically viable. Prior to its development, it is essential to perform a cost-benefit analysis of the strategy implementation. Similar analyses—completed in other countries (specify) using varying methodologies—have shown that the implementation of improved stoves is viable [6,7].

This paper reinforces the conclusion of the feasibility of technology presented in [6], but using a different methodology and the assumptions of a National Technology Adoption Strategy.

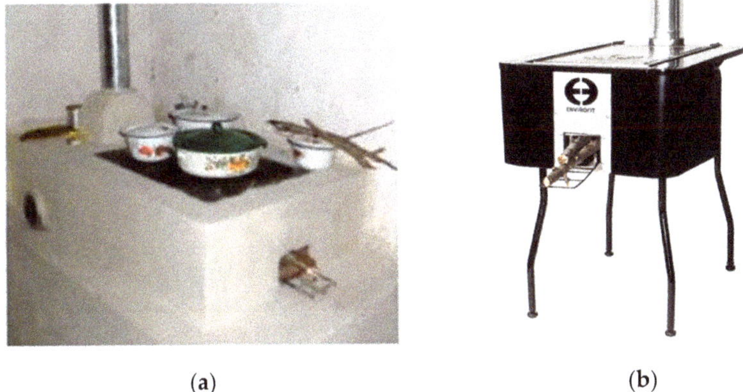

(a) (b)

Figure 2. Traditional and improved stoves used for cooking in Honduras. (https://envirofit.org/honduras/). (**a**) Traditional firewood stove (**b**) Improved stove.

Thus, the methodology implemented included a review of literature and interviews with the stakeholders of the improved stoves value chain in Honduras. For the cost-benefit analysis, the *Long-range Energy Alternatives Planning System*, LEAP® (Software version: 2018.1.37, Stockholm Environment Institute. Somerville, MA, USA) tool was used [8]. This tool is widely used in the analysis and formulation of energy policies and strategies worldwide. This tool considers the demand for firewood, as well as the gradual introduction of improved stoves for cooking food, according to the assumptions of a National Technology Adoption Strategy.

2. Material and Methods

2.1. Current Status of Improved Stoves Delivery for a National Strategy Adoption in Honduras

This subsection presents the stakeholders, projects, and the NAMA (Nationally Appropriate Mitigation Actions) program according to a national strategy for the adoption of improved stoves in Honduras.

2.1.1. Stakeholders and Projects

The companies dedicated to the promotion and construction of improved stoves are currently small, non-profit or growing social enterprises with minimal capital, which basically depend on sales through contracts signed with non-governmental organizations, who in turn depend mainly on donations from small local or international initiatives. There is neither a wide market for improved stoves, nor any chance of one being generated if the state continues giving away the stoves [9,10].

A case to highlight this concept is the Mirador project, which finances part of its activities using carbon credits [11,12]. Putting an experience into practice under this certification process is costly. Alternatively, it is different from other initiatives due to its funding source, which has a component to monitor and evaluate the installation of improved stoves [12].

In recent years, joint efforts have been made in order to coordinate activities and strengthen the value chain of improved stoves. The Government of Honduras (GoH), along with international cooperators, academics, and the private sector, has participated [6,7] in these efforts. Figure 3 shows the relationships of some stakeholders, as well as other agents, currently present in the delivery of improved stoves.

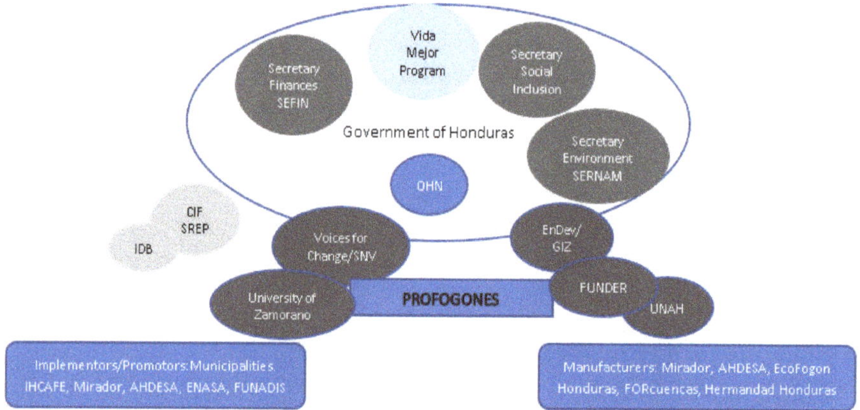

Figure 3. Stakeholder mapping of the clean cookstove sector in Honduras (modified from [9]).

Programs for the production, distribution and adoption of improved stoves in Honduras date back to the end of the last century; however, their greatest period of momentum has occurred in the current decade. International organizations—together with Honduran non-governmental organizations—initiated small scale programs during the past two decades [7,9]. These programs proved the advantages to health, forest conservation and energy efficiency when traditional stoves were replaced with improved ones.

The Honduran government joined these efforts in 2013, with a comprehensive manufacturing and distribution program, titled the *Better Life Program* [7]. Up to the end of 2017, around 600,000 improved stoves had been distributed throughout the country [7] (see Table 1). However, this number does not necessarily mean the stoves are currently being used, as not all people who received them have adopted the technology as of yet [13].

Table 1. Producers and improved stoves installed to December 2017 (Data from [7]).

No	Producer	Quantity of Improved Stoves	% of Share
1	Enviroeit (GoH)	256,679	44.0
2	Proyecto Mirador	170,767	29.3
3	Ahdesa	50,000	8.6
4	Fundeih (GoH)	34,407	5.9
5	EnDev/Focaep	33,000	5.7
6	Profogones	11,346	1.9
7	Proparque	7404	1.3
8	Ornader	6030	1.0
9	Funda Ahprocafe	6000	1.0
10	Gema/Usaid	4270	0.7
11	Acceso	2240	0.4
12	Clifor	1152	0.2
	Total	583,295	100

There are only a small number of commercial suppliers that sell improved stoves in the country. Table 1 shows that the majority of suppliers are programs and Non-Governmental Organizations (NGOs). The GoH stands out with a 44% share in the production and distribution of stoves through the Envirofit and Fundeih (Envirofit Honduras and Fundeih are part of "Vida Mejor" Goverment Program. Envirofit build the stove and the Goverment pay to Fundeih, which distributes the stoves.) programs since 2013. The second largest program is Mirador project, an NGO that has been working in Honduras since 2004 and that has distributed about 180,000 improved stoves (equivalent to 29.3% of the total). Additionally, Adehsa, Fundeih and Endev/Focae are also suppliers, with shares of 8.6%, 5.9% and 5.7%, respectively. Other smaller programs are also participating [7].

The goals and characteristics of these programs have not been homogeneous, although all are based on the benefits of replacing the traditional stove with an improved one. The main difference is whether objectives include the creation or expansion of the market for improved stoves. There are three market segments identified: (1) families in extreme poverty that are not able to pay for an improved stove and therefore require a total subsidy; (2) a second segment of limited economic capacity that requires a partial subsidy; (3) a third segment that operates in the free market of improved stoves.

Hence, for the first segment, programs should be aimed at those in extreme poverty; in such cases the improved stove would be donated. On the other hand, the Mirador project, although highly subsidized, also requires local inputs in terms of materials and labor [7]; this would be the case with the second segment. The program EnDev/Focaep seeks to create a market for improved stoves through paying attention to the different components of the value chain. In the same way, the Profogones project promotes a sustainable business model for improved stoves. The latter is linked to the *Vida Foundation*, with the Inter-American Development Bank (IDB) as the project administrator.

In practice, these programs could be considered complementary, due to the market segment they seek to fulfill. However, the way in which the government program is executed—i.e., with political objectives—distorts the rest of the market segments.

2.1.2. Nationally Appropriate Mitigation Actions (NAMA)

Another effort to coordinate actions is Nationally Appropriate Mitigation Actions (NAMA), the objective of which is to increase the adoption of improved stoves in low-income households in Honduras. One of the main goals of NAMA is to bring improved stoves to 1.126.000 families by 2030 [10]. In the same way, NAMA will promote coordination and communication among stakeholders, generating comparable and transparent information, as well as the contributing to a common report of national advances in the reduction in greenhouse gases.

On the other hand, NAMA can also contribute to the strengthening of micro, small and medium enterprises that manufacture improved stoves and to the supply chain, due to the increased demand in the market.

Considering the need to unify and create synergies among multiple initiatives, the coordination of stakeholders and various programs of improved stoves will be one of the main challenges for NAMA and the National Strategy. Therefore, it is proposed that a *National Bureau of Improved Stoves*—that will benefit the coordination of the different stakeholders in NAMA—is established [10].

2.2. Methodology and Data Used in the Cost-Benefit Analysis of a Strategy for Adoption of Improved Stoves in Honduras

The methodology used to evaluate the cost-benefit of implementing a National Strategy for the adoption of improved stoves is based on using the LEAP (Software version: 2018.1.37, Stockholm Environment Institute. Somerville, MA, USA) software.

LEAP is an integrated, scenario-based modeling tool that can be used to track energy consumption, production, and resource extraction in all sectors of an economy. It can be used to account for both the energy sector and the non-energy sector, as well as greenhouse gas emission sources and sinks. In addition, LEAP can also be used to analyze emissions of local and regional air pollutants and short-lived climate pollutants, making it well-suited to studies of the climate co-benefits of local air pollution reduction [4,8].

LEAP is not a model of any particular energy system, but rather a tool that can be used to create models of different energy systems, in which each requires its own unique data structure. LEAP supports a wide range of modeling methodologies [6]. On the demand side, these range from bottom-up, end-use accounting techniques, to top-down macroeconomic modeling [8].

LEAP's modeling capabilities operate at two basic conceptual levels. At one level, LEAP's built-in calculations handle all the "non-controversial" energy, emissions and cost-benefit accounting calculations [8]. At the second level, users enter spreadsheet-like expressions that can be used to specify time-varying data or to create a wide variety of sophisticated multi-variable models, thus enabling econometric and simulation approaches to be embedded within LEAP's overall accounting framework [8].

In this study, LEAP is used for the calculation of the costs and benefits of implementing a strategy for the adoption of improved stoves in the urban residential sector (electrified and non-electrified), the rural sector and the commercial sector, with and without shares of Liquefied Petroleum Gas (LPG). The base year is 2016, and the target year is 2030. Variables were also established to be the most representative for the analysis of the energy sector: Population, GDP, income, households, GDP growth, population growth and demand growth.

According to the 2016 Honduras Energy Balance, the final energy consumption is 56.33% primary energy and 43.67% secondary. The final consumption of primary energy was divided into the main consumption sectors—residential, commercial and industrial. The share of each sector of primary energy consumption was determined as follows: the industrial sector with 13.17% energy consumption share, the commercial sector with 4.76% share, and the residential sector with 82.07% share. The latter value represents majority of the share.

The residential area was divided into urban and rural areas with shares of 54.1% and 45.9% of energy consumption, respectively. This energy consumption is driven by the factors of both rising household quantities and rising population.

Therefore, for both areas previously mentioned, the firewood consumption was taken. For the urban residential sector, 25% of households consume firewood, and for the rural residential sector, 77.96% consume firewood.

It is established that the traditional stoves account for an approximate 7.45 m^3 yearly consumption of wood per household, and the improved stoves accounts for only 2.13 m^3 per household.

For secondary energy consumption in the residential sector, the sector was divided into urban and rural areas, and each of these areas was classified into electrified and non-electrified.

Electrified zones use mainly lighting, cooling, and cooking. In the cooking section, LPG was added, which represents 42% of the energy used for cooking; an average consumption value of 300 pound per year was assumed considering that a 25-pound container is consumed in each home per month.

On the other hand, by considering historic consumption, it is assumed that under reference scenario the LPG consumption per households will grow 18.4% per year.

For the non-electrified area, only the kerosene for lighting and the LPG for cooking are considered. In this scenario, only the LPG consumption for food cooking is analyzed, mainly in the peri-urban area of Tegucigalpa, the capital of Honduras. In this category, the use of LPG will rise to 36.8% in 2030. This is due to an assumed National Policy by the GoH, aimed to encourage the use of LPG due to the increasing electricity tariff. Finally, it is considered that there will be no increase in the use of LPG in rural areas.

2.2.1. Scenarios

Three scenarios were used in the analysis, as follows:

- Business as Usual (BAU)—a scenario in which the strategy is not implemented. This scenario does not consider the implementation of measures to adopt the new technology. Under this scenario, the government continues giving away the improved stove as it was mentioned in the previous section.
- The scenario with a strategy. Under this scenario, improved stoves are introduced in the urban and rural households.
- The final scenario analyzed is the introduction of improved stoves plus LPG.

By 2017, 583,295 improved stoves had been delivered, of which 20% have not been adopted by users (116,659 stoves). It is expected that by 2030, 1,125,000 improved stoves will have been already been installed, which implies that 658,364 improved stoves should be installed in that time.

2.2.2. Manufacture Costs

The manufacturing costs of improved stoves are as follows:

- Urban households: *Justa* portable stove, USD 61.78.
- Urban households: *Justa* 2 × 3 stove, USD 59.50.
- Commercial: *Justa* 22 × 22 stove with flatiron, USD 108.16.

These costs are introduced into the LEAP model, in such a way that they were annualized throughout the analysis period. Thus, the following figures (Figure 4, Figure 5, Figure 6) were obtained, which show the costs behavior from the base year up to 2030. It is assumed that a traditional stove has a cost of USD 34.00.

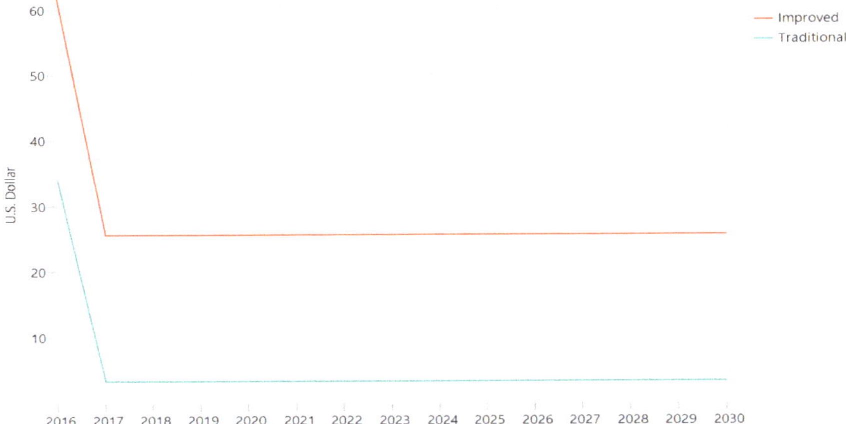

Figure 4. Annualized cost of improved stoves for urban households.

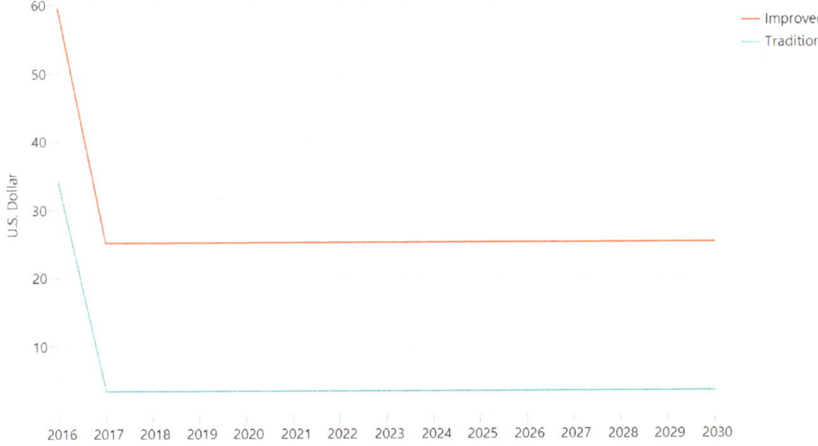

Figure 5. Annualized cost of improved stoves for rural households.

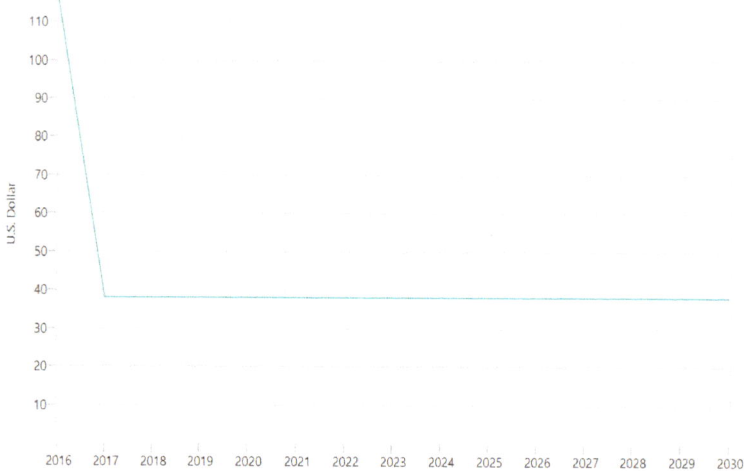

Figure 6. Annualized cost of improved stoves for the commercial sector.

On the other hand, the benefits of implementing a strategy for improved stove-adoption are broadly known:

- The improvement of air quality—a reduction in particulate emissions (black carbon) and smoke.
- Reduction in fuel needs (saving time and money), particularly benefiting women and children who traditionally collect firewood.
- The creation of new jobs in production, sales, marketing and distribution of improved stoves.
- Reduction in pressure on the forest.
- Health benefits as a result of the reductions in household air pollution.
- Others.

Furthermore, before analyzing the cost-benefit of each scenario in comparison with the reference scenario, it is important to observe the energy consumption behavior of each scenario and contrast that behavior with the reference scenario, in order to have a better idea of what the implication of energy use in the cost-benefit analysis is.

Hence, the results of the energy consumption dynamics of each scenario are shown first. Then, the results of the cost-benefit analysis are presented.

3. Strategy Implementation Results

3.1. BAU Scenario

As mentioned earlier, in this scenario, the same considerations are being made under the same procedures throughout the study period. Figure 7 shows the household growth in Honduras up to 2030. This growth is 2.62% per year, according to official data.

Figure 8 shows that under the BAU scenario, energy consumption is constantly growing throughout the analysis. This figure only shows the primary energy consumption, which in this analysis considers solely firewood and bagasse. Bagasse is used in industrial demand, but this is not subject to the analysis for the implementation of an improved stoves strategy in energy demand, mainly for cooking food.

Figure 9 shows that the implementation of improved stoves in urban areas would follow a slow growth throughout the analysis period. Under this scenario, traditional stoves would be the main energy source needed for cooking food. Such stoves are based on burning firewood. The same behavior in energy consumption is shown in the rural area, as depicted in Figure 10. However, in rural areas, firewood consumption is higher.

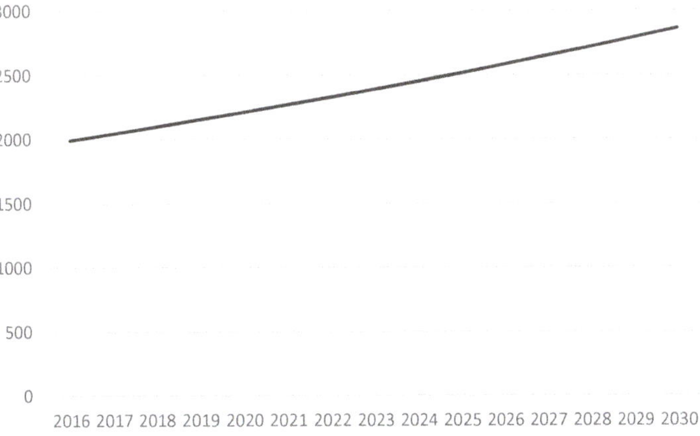

Figure 7. Household growth (Thousands of households per year).

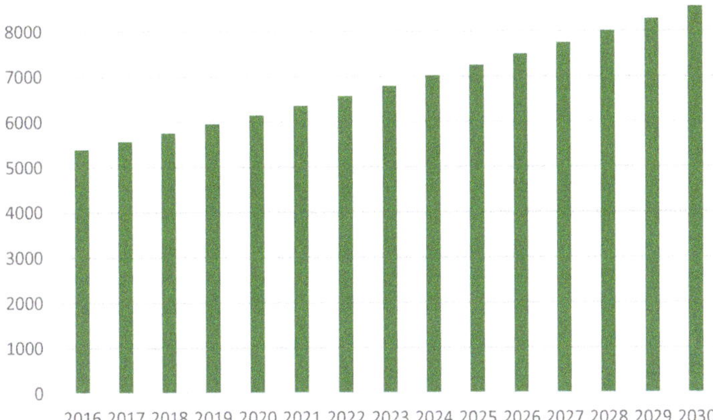

Figure 8. Total primary energy demanded under the BAU scenario (Thousands of Barrel of Oil Equivalent per year).

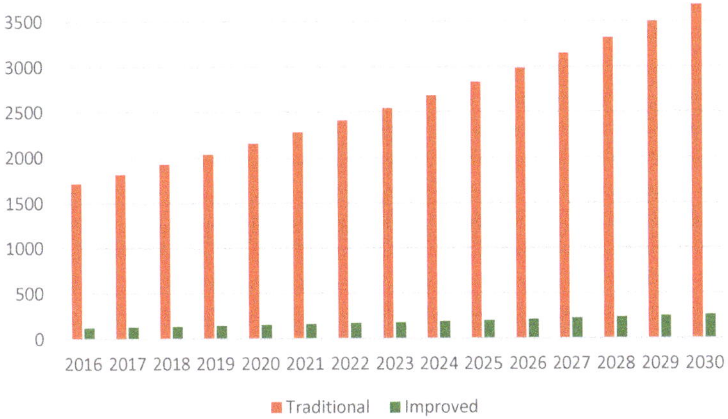

Figure 9. Firewood demand in urban households under the BAU scenario. (Thousands of Barrel of Oil Equivalent per year).

Figure 10. Firewood demand in rural households under the BAU scenario. (Thousands of Barrel of Oil Equivalent per year).

3.2. Introduction of Improved Stoves vs. BAU Reference Scenario

Under this scenario, the introduction of improved stoves in the Honduran energy sector is analyzed according to a National Strategy, whose goal is the installation and adoption of 1,125,000 improved stoves for cooking food.

Figure 11 shows that for the urban residential sector, the sharing of improved stoves implies a lower energy consumption throughout the analyzed period, in relation to the reference scenario (bars without color). In the same way, it is shown that traditional stoves should reduce their share at the end of the same period.

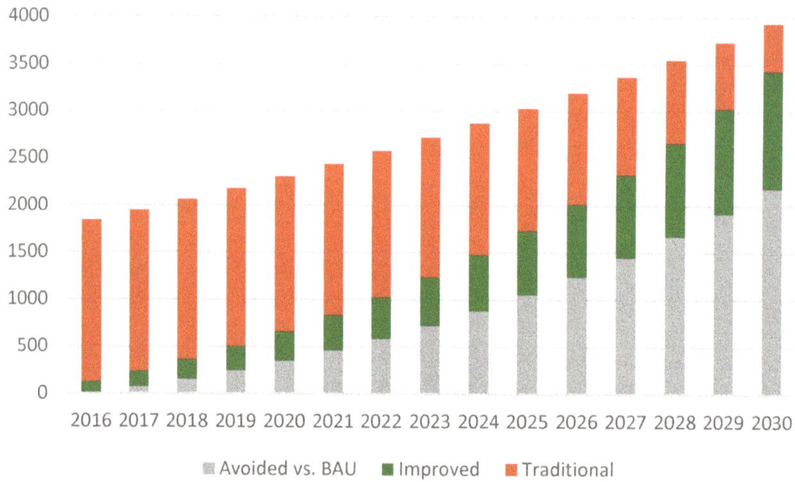

Figure 11. Firewood demand in the urban area according to the annual introduction of improved stoves until 2030. (Thousands of Barrel of Oil Equivalent per year).

Figure 12 shows that for rural areas, the energy avoided (bars without color) is less than for urban areas. However, the introduction of improved stoves decreases energy consumption throughout the analyzed period. This makes the sector more efficient in terms of the consumption of primary energy (firewood). It should be noted that when observing the scales in both figures, more wood is consumed in the rural area. The latter is verified by observing Figure 13, which shows the consumption of firewood for the urban and rural areas, considering both improved and traditional stoves.

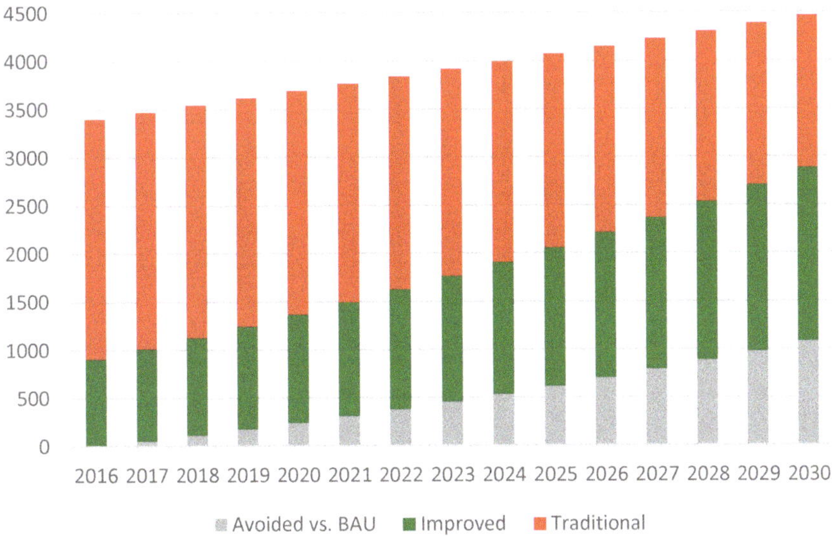

Figure 12. Firewood demand in the rural residential area according to the annual introduction of improved stoves until 2030. (Thousands of Barrel of Oil Equivalent per year).

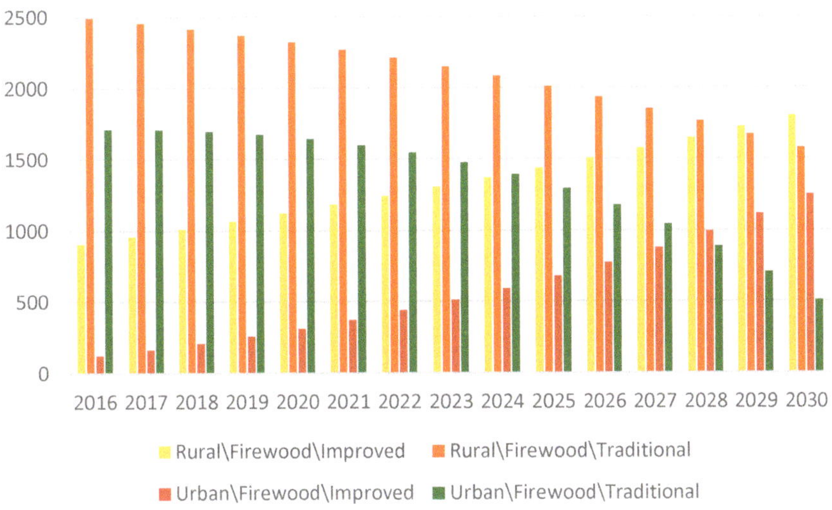

Figure 13. Energy demand in both urban and rural residential areas according to the annual introduction of improved stoves until 2030. (Thousands of Barrel of Oil Equivalent per year)

Figure 14 shows that if improved stoves are introduced in the commercial sector under this scenario, the consumption of firewood would be reduced throughout the analyzed period. For that reason, 22,000 barrels of oil (BEP) would be avoided—and that is only by 2030.

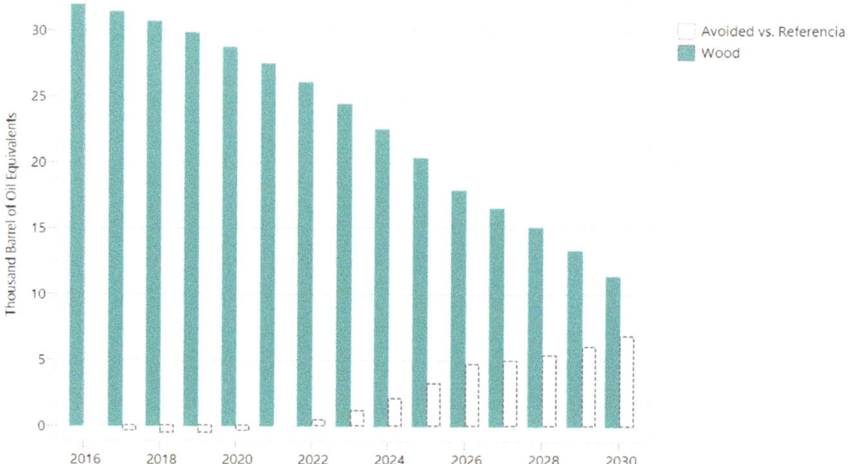

Figure 14. Firewood demand in commercial sector vs. what would be avoided according to BAU scenario.

3.3. *Introduction of Improved Stoves and LPG vs. BAU Scenario*

Figure 15 shows that under this scenario, LPG consumption increases throughout the analysis period. This observation is noticeable for the urban, electrified and non-electrified residential areas, as well as for the rural electrified households. These results are consistent with the fact LPG consumption will increase in the peri-urban areas of the urban sector.

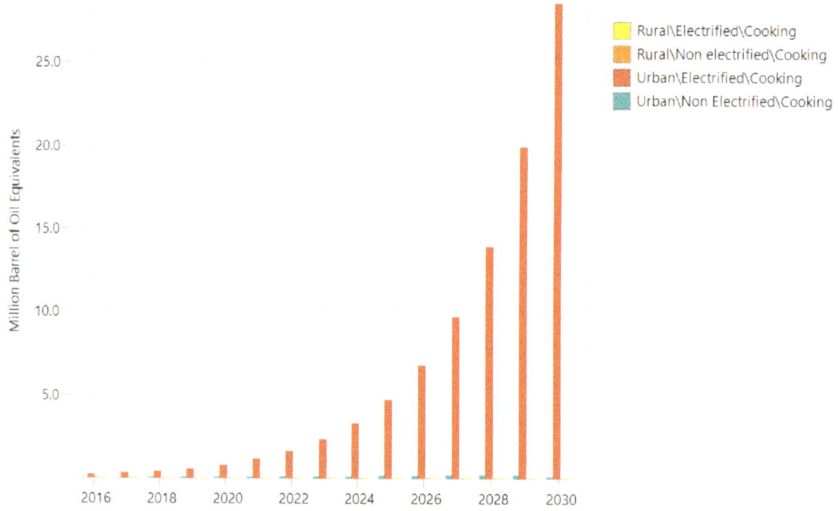

Figure 15. LPG consumption for the stoves plus LPG scenario.

On the other hand, Figure 16 shows that in rural, non-electrified areas, it is expected that the consumption will be reduced even more. This due to the rise consumption of firewood.

Figures 17 and 18 show that more LPG is consumed under this scenario, both in the urban electrified and non-electrified areas. The label "all others" represent the years before 2021.

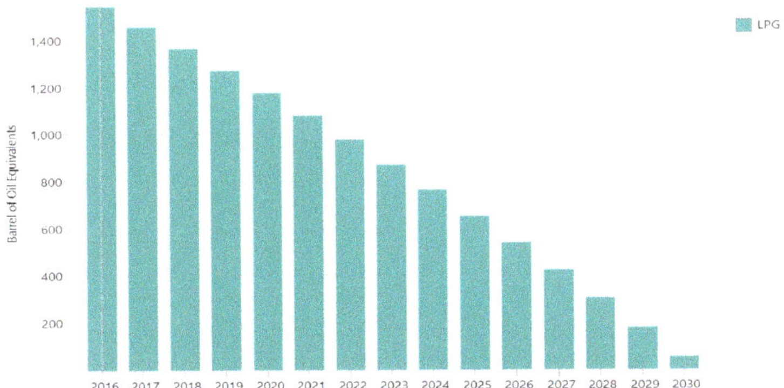

Figure 16. LPG consumption for the stoves plus LPG scenario. Rural residential area without access to electricity.

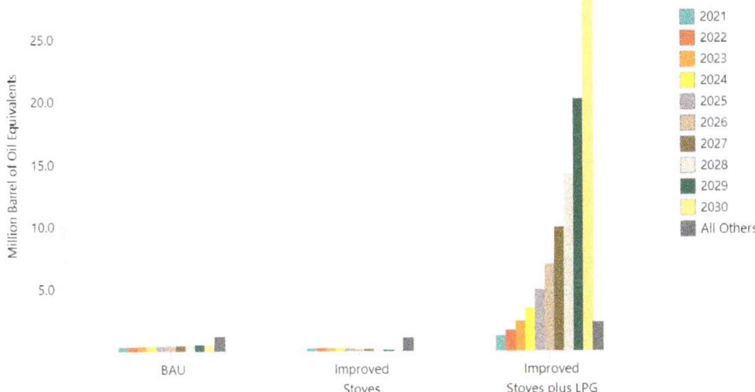

Figure 17. Comparison of the different scenarios in the LPG consumption for the stoves plus LPG scenario. Period 2021–2030. Electrified urban residential area.

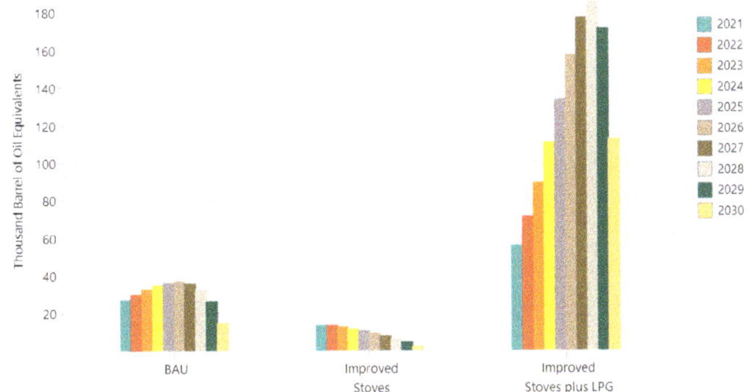

Figure 18. Comparison of the different scenarios in the LPG Consumption for the stoves plus LPG scenario. Period 2021–2030. Urban residential area not electrified.

3.4. *Environmental Burden for the Different Scenarios*

The following figures show the emissions observed in the different scenarios. According to Figures 19 and 20, emissions resulting from a BAU reference scenario are greater than a scenario under which a strategy of "Introduction of Improved Stoves" is implemented. On the other hand, under the scenario of LPG and improved stoves, emissions are higher (see Figure 21) than the emissions from the BAU scenario.

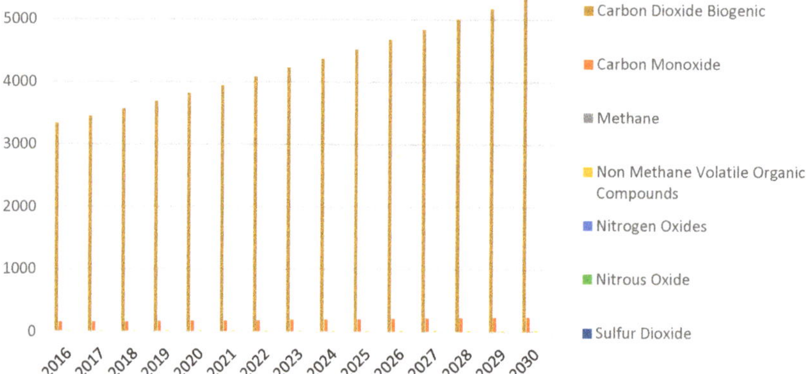

Figure 19. Emissions under the BAU scenario. (Thousands of Metric Tonnes)

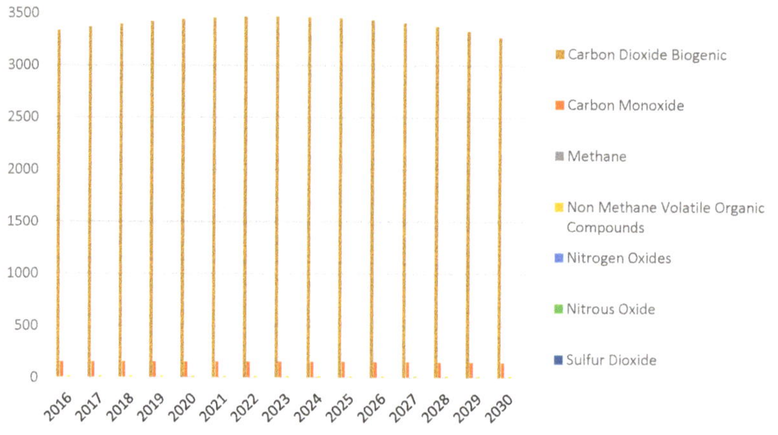

Figure 20. Emissions under the improved stoves scenario. (Thousands of Metric Tonnes).

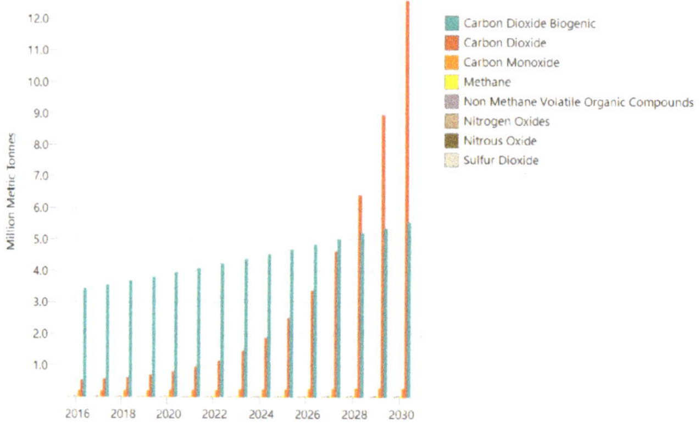

Figure 21. Emissions under LPG and improved stoves scenario.

3.5. Cost-Benefit of the Implementation of a Strategy for the Adoption of Improved Stoves in Honduras

The LEAP tool shows that the improved stoves scenario is cheaper than the reference scenario (Table 2). This is concluded from the Net Present Value, which for the improved stoves scenario is USD 1253.8 million cheaper than the BAU scenario. For this reason, it would be cheaper to implement an improved stove-adoption strategy in the Honduran energy sector than to not. This conclusion includes the direct manufacturing costs of improved stoves as well as the costs of firewood for cooking.

Table 2. Cumulative Costs and Benefits of an Improved Stoves Strategy in Honduras: 2016–2030. Relative to Scenario: BAU. Discounted at 5.0% to year 2016. (Units: Million 2016 U.S. Dollar).

	Improved Stoves	LPG Plus Improved Stoves
Demand	−1253.8	376.7
Primary Energy	−844.3	−185.9
Secondary Energy	−409.5	562.6
Net Present Value	−1253.8	376.7
GHG Savings (Mill Tonnes CO_2e)	2.5	−38.8
Cost of Avoiding GHGs (U.S. Dollar/Tonne CO_2e)	−496.7	

The cost of implementing such a strategy, considering the consumption of firewood (primary energy), is USD 844.3 million cheaper than the strategy's non-implementation.

On the other hand, the LPG plus improved stoves scenario shows a positive net present value of USD 376.7 million, so this scenario is more expensive than the reference scenario. The reason for this is that the share of LPG implies the import of a fuel that is not produced in the country.

Furthermore, the cost of avoiding emissions is lower in the scenario of improved stoves, at USD 496.7 per ton, in relation to the reference scenario. Hence, the implementation of an "Improved Stoves Strategy" in Honduras would reduce the emission of greenhouse gases more economically than the non-implementation of the strategy.

4. Discussion: Towards a National Strategy for the Adoption of Improved Stoves

Despite the existence of the structure showed in Section 2.1, strong leadership is necessary to achieve the objectives related to the support of the value chain in the process of adopting improved stoves.

Thus, the design and execution of a National Strategy for the adoption of improved cookstoves requires an institutional framework that considers not only the progress obtained so far, but also the challenges of the future. This requires leadership that actively promotes the different components of the strategy with a long-term vision. Therefore, such an integral policy should be implemented under the leadership of the GoH, given the need to coordinate efforts with different stakeholders.

Hence, among the different components for a National Strategy, the following must be included:

a. National Standard for Improved Stoves

When Honduras officially launched the standard of improved stoves OHN 97001.2017 [11], as part of the *PROFOGONES* project, the country became the third country in Latin America to establish the performance requirements to categorize improved stoves. The implementation of this standard promotes the dissemination of improved stoves for sustainable development in terms of health for users, reduction of pollutant emissions, an adequate use of natural resources, and economic benefits for users.

The OHN 97001:2017 standard establishes the minimum requirements of efficiency, safety, and quantity of intra-household emissions captured from an improved stove by categorizing models according to their performance.

b. Training Programs to Improve the use of Efficient Stoves and the Efficient Use of Firewood

One of the main goals of the National Strategy must be to make users aware of the benefits of using improved stoves. Training is important, as when the potential users are aware of the damages and ailments caused by smoke derived from the use of firewood, they will be able to better understand the need to change their method of cooking. This technological change implies strong behavioral changes regarding fuel, technology, and cooking; therefore, it is necessary to accompany users in this process, so that they do not abandon the technology in the face of difficulties [12].

c. Promotion of Financing Mechanisms

Evidence obtained during this study in Honduras shows that it is better to have an open market, stratify the target population who will be involved, know the material benefits, consider the subsidy according to the stratification of the participating population, and boost a market of pieces and parts of improved cookstoves. Evidence obtained during this study shows that it is better to have an open market, stratify the target population and subsidies, know the material benefits, and boost a market for the pieces and parts of improved cookstoves. Families unable to pay the total cost of an improved stove could be asked to cover a part of the cost working in the installation process. This participation improves the adoption of the new technology.

For the user who can pay, financing mechanisms must be created through local and/or regional credit institutions, i.e., rural savings banks, cooperatives, among others.

d. Monitoring and Evaluation

Currently, most programs that promote the establishment of improved stoves in Honduras are measured by the number of stoves built, distributed, and/or sold. However, this does not mean that the technology has been adopted and stoves are effectively being used. Few programs carry out monitoring and evaluation [14,15]. Therefore, in a National Strategy, it is important to broaden the approach of evaluating the process of building, distributing, selling and adopting stoves, to a methodology that includes the monitoring and evaluation of their use as well.

e. Certification and Applied Research

The certification will be used to evaluate the different types of stoves based on three characteristics established by the Honduran OHN 97001 standard for improved stoves [16]: (1) reduction in fuel use, (2) the capacity to reduce emissions, and (3) user safety. The foregoing will ensure that all stoves that are put into service meet the minimum standard criteria of fuel efficiency, indoor air quality, particles emissions and carbon monoxide, durability, and safety.

f. Stove Users and Producers' Associations

The main stockholders to consider will be users from low-income households in urban and rural areas that use firewood with traditional stoves. Women and children are the most exposed to air pollution inside the house. For this reason, female leaders must be trained in rural communities and neighborhoods in peri-urban areas as promoters responsible for coordinating demand and monitoring. Similarly, the training of master builders, i.e., builders of improved stoves, is needed.

5. Conclusions and Policy Implications

The cost-benefit analysis for the implementation of an Improved Stoves Strategy in Honduras was performed using the *Long-range Energy Alternatives Planning System* (LEAP) tool. The model shows the following results:

- A strategy for the introduction of improved stoves benefits the energy sector, since the consumption of firewood would be reduced.
- Implementation of an improved stoves strategy would be cheaper than continuing with the current scenario.

- ○ The cost of avoiding emissions is lower if an improved stove strategy is implemented, compared to continuing with the current scenario of improved stove delivery.

There are many stakeholders interested in the value chain of improved stoves in Honduras, a strategy for the adoption of this technology would have an impact on the process improvement and a reduction in direct costs and environmental externalities.

On the other hand, some lessons learned in the process of manufacturing and delivering improved stoves in Honduras could be the following:

- It is necessary to have an institutional leader in order to obtain improved results.
- Funds used in these projects must be clearly audited.
- Rural cooperatives have shown good performances in the manufacture and distribution of improved stoves in Honduras.
- In order to create value in the manufacturing process, the manufacturer must be trained.
- Different universities and educational institutes must be involved to improve the research and development process.

Finally, the economic valuation of the external environmental benefits is difficult under this project. However, the authors believe this could be a good opportunity for future research in this important field of study.

Author Contributions: Conceptualization, W.C.F., B.B. and H.N.P.; methodology, W.C.F.; validation, W.C.F., B.B. and H.N.P.; formal analysis, W.C.F., B.B. and H.N.P.; investigation, W.C.F. and B.B.; resources, W.C.F., B.B. and H.N.P.; data curation, W.C.F.; writing—original draft preparation, W.C.F., H.N.P., S.R. and A.A.-S..; writing—review and editing, W.C.F., B.B., H.N.P., S.R. and A.A.-S.; visualization, W.C.F. and S.R.; funding acquisition, H.N.P., S.R. and A.A.-S. All authors have read and agreed to the published version of the manuscript.

Funding: Ministry of Foreign Affairs of The Netherlands and Khalifa University.

Acknowledgments: This study was possible through funding of the Ministry of Foreign Affairs of The Netherlands, through the Voice for Change Partnership program lead by SNV Netherlands Development Organization. Furthermore, it is supported by Khalifa University under Award FSU-2018-25.

Conflicts of Interest: The authors declare no conflict of interest.

References

1. INE, National Institute of Statistics (Honduras). Available online: https://www.ine.gob.hn (accessed on 14 February 2020).
2. CESPAD. La Deforestación del Bosque en Honduras: Entre Tala Ilegal y una Endeble Institucionalidad. 2015. Available online: http://cespad.org.hn/wp-content/uploads/2017/06/Monitoreo-RRNN-oct-2.pdf (accessed on 14 February 2020).
3. Jebaraj, S.; Srinivasa, P. High-Efficiency Solar Oven for Tropical Countries, ARPN. *J. Eng. Appl. Sci.* **2015**, *10*, 10213–10317.
4. Flores, W.; Ojeda, O.; Flores, M.; Rivas, F. Sustainable energy policy in Honduras: Diagnosis and challenges. *Energy Policy* **2011**, *39*, 551–562. [CrossRef]
5. Wilfredo, F.C. *El Sector Energía de Honduras: Aspectos Necesarios Para su Comprensioón y Estudio*; Library of U.S. Congress: Comayaguela, Honduras, 2016; ISBN 978-99926-52-78-7. Available online: https://lccn.loc.gov/2018407916 (accessed on 14 February 2020).
6. García-Frapolli, E.; Schilmann, A.; Berrueta, V.; Riojas-Rodríguez, H.; Edwards, R.; Johnson, M.; Guevara-Sanginés, A.; Armendariz, C.; Masera, O. Beyond fuelwood savings: Valuing the economic benefits of introducing improved biomass cookstoves in the Purépecha region of Mexico. *Ecol. Econ.* **2010**, *69*, 2598–2605. [CrossRef]
7. Pino, H.; Benjamin, B. *Estudio del Impacto de la Exoneración del Impuesto Sobre Venta a las Estufas Mejoradas y el Impuesto de Importación a las Láminas y Otras Partes Importadas para la Fón de Estufas Mejoradas en Honduras*; Voz para el Cambio, SNV: Tegucigalpa, Honduras, 2018.
8. Heaps, C.G. *Long-range Energy Alternatives Planning (LEAP) System*; Software Version: 2018.1.27; Stockholm Environment Institute: Somerville, MA, USA, 2016.

9. GDI. *Promoting Sustainable Business Models for Clean Cookstoves Dissemination in Honduras, Global Delivery Initiative*; GDI, 2018. Available online: https://globaldeliveryinitiative.org/sites/default/files/case-studies/cif-gdi-_honduras_case_study_-final.pdf (accessed on 20 December 2019).
10. Ludeña, C.E.; Salomon, M.; Cocco, M.; Dannecker, C.; Grütter, J.; Zelaya, S. *Identificación y priorización de Acciones Nacionales Apropiadas de Mitigación (NAMA) en los sectores de agricultura, transporte y eco-fogones en Honduras*; Banco Interamericano de Desarrollo: Washington, DC, USA, 2015.
11. OHN 97001:2017. *Estufas Mejoradas—Requisitos y Métodos de Ensayo para la Clasificación y Categorización*; OHN: Tegucigalpa, Honduras, 2017; Organismo Hondureño de Normalización.
12. OLADE. Uso Racional y Sostenible de la Leña en los Países de SICA. 2013. Available online: http://www.olade.org/wp-content/uploads/2015/08/UsoLe%C3%B1a_OLADE-SICA-2013.pdf (accessed on 20 December 2019).
13. Minister of Environment and Natural Resources of Mexico. Comparative Study of Improved Stoves to Sustain an Intervention Program Massive in Mexico. September 2009. Available online: https://www.cleancookingalliance.org/resources_files/estudio-comparativo-de.pdf (accessed on 29 December 2019).
14. Hafner, J.; Magenau, H.; Uckert, G.; Sieber, S.; Graef, F.; König, H. Four years of sustainability impact assessments accompanying the implementation of improved cooking stoves in Tanzania. *Environ. Impact Assess. Rev.* **2020**, *80*, 221–236. [CrossRef]
15. Adrianzén, M. Improved cooking stoves and firewood consumption: Quasi-experimental evidence from the Northern Peruvian Andes. *Ecol. Econ.* **2013**, *89*, 135–143. [CrossRef]
16. Kedir, M.; Bekele, T.; Feleke, S. Problems of Mirt, and potentials of improved Gonzie and traditional open cook stoves in biomass consumption and end use emission in rural wooden houses of Southern Ethiopia. *Sci. Afr.* **2019**, *3*, e00064.

Article

Public Perceptions of Energy Scarcity and Support for New Energy Technologies: A Western U.S. Case Study

Alexandra Buylova [1,*], Brent S. Steel [1] and Christopher A. Simon [2]

[1] School of Public Policy, Oregon State University, Corvallis, OR 97331, USA; bsteel@oregonstate.edu
[2] Department of Political Science, University of Utah, Salt Lake City, UT 84112, USA; simon@cppa.utah.edu
* Correspondence: buylovaa@oregonstate.edu; Tel.: +1-541-737-2811

Received: 2 December 2019; Accepted: 28 December 2019; Published: 3 January 2020

Abstract: This study examines public concern for energy security and support for public investment in new energy technologies. Using household survey data from the western U.S. states of California, Idaho, Oregon, and Washington, socio-demographic characteristics, environmental values, and policy relevant knowledge are analyzed as drivers of energy security and technology investment orientations. Findings suggest that a majority of respondents in each state believe that not enough money is being spent on energy research, that the country has insufficient energy resources, and that new technologies can support future energy security. Multivariate analyses indicate that some socio-demographic variables (e.g., gender and education), ideology, and environmental value orientations also have an impact on energy security orientations and support for technology investment.

Keywords: energy technology; energy security; public opinion

---

1. Introduction

This study contributes to the literature on determinants of public perceptions of new energy technologies and energy security by analyzing the impact of public energy-related knowledge, environmental value orientations, political ideology, and socio-demographic characteristics on public perceptions of energy security and new energy technologies. More specifically, using public opinion survey data from four Western states in the United States, we investigate public perceptions of: (1) the state of the country's energy supply; (2) being personally affected by the shortage of electricity in the next 5 years; (3) support for government investment into new energy technologies; and (4) the ability of new energy technologies to meet future energy demands.

In the process of policy formulation and implementation, one cannot ignore public opinion, especially in democratic societies like the United States [1–5]. Motivation to investigate public opinion toward energy-related issues and new energy technologies comes from the fact that the U.S. is a high energy consumer society with heavy reliance on fossil fuels in electricity generation and the transportation sector. Therefore, energy supply security is a central political and policy issue. At the same time, the country strives toward a low-carbon economy, diversifying its energy portfolio to include a larger share of renewable energy and other alternative energy technologies, including smart meters, electric vehicles, carbon capture, storage, and energy efficiency technologies [6]. Such policy innovations reflect the country's planning of energy independence and security, where renewables (i.e., wind, sun, biomass, nuclear) can be an alternative to traditional energy sources (e.g., coal, oil, and gas), which are finite in supply and are influenced by global fuel market price fluctuations [7–9]. A number of studies find strong support among the general public for renewable energy as a major source for future electricity portfolios [10–13]. In addition, the transition to low-carbon sources of energy satisfies environmental concerns and provides the added benefit of reduced marginal social costs, allowing the U.S. to respond to international diplomatic pressures of reducing CO_2 emissions from burning conventional fossil fuels. Despite a number of climate commitments, the U.S. remains one of the top emitters of greenhouse gases per capita [14,15].

Extant research on public support and opposition toward new energy technologies emphasizes the role of place, geographic proximity, land-use regulations, socio-economic impacts, fairness, and trust in shaping public opinion [2,16–18]. These studies investigate public perceptions of concrete energy projects, which are likely to carry specific drawbacks or opportunities for communities directly affected by new developments. On a more abstract level, other studies explore the general public's familiarity with new energy technologies, including wind energy [19], offshore renewable energy [20], smart meters [21,22], and electric vehicles [23]. These studies find that public opinion concerning energy technologies is often rooted in the degree to which those technologies are perceived as risky, with uncontrollable and catastrophic impacts [24], or tampering with natural processes [25–27].

However, to understand public opinion on broader energy policies in the era of low-carbon energy transition, there is a need to further analyze public orientations on energy-related questions. There is a lack of research that inquires into public perceptions of national and personal energy security issues, the level of government funding towards new energy technologies, and the ability of new technologies to meet energy demands of the future. This research addresses this gap in the literature. Moreover, we contribute to research on public opinion of energy policy and technologies by investigating the drivers of public perceptions that include environmental value orientations, political ideology, public knowledge, and socio-demographic characteristics.

1.1. Environmental Value Orientations and Ideology

Environmental values are commonly measured utilizing the New Ecological Paradigm (NEP) from Dunlap et al. [28]. The NEP scale consists of a range of ecological worldview aspects, such as a personal stance on humans' place in the ecosystem, the balance of nature, the rights of humans to modify the environment, and others. As expressed in the Values-Beliefs-Norms (VBN) model of environmental decision-making [29], values, or intuitive rather than calculative logic, can serve as reliable indicators of perceptions toward emergent clean energy technologies [30]. Extant literature finds support for the pro-NEP position as a significant indicator of positive attitudes toward new alternative energy sources [31] and government investments in alternative energy [32]. In this study, we investigate if and in what way environmental values shape public perceptions of new energy technologies, government investments in new energy technologies, and concerns regarding the security of energy supplies.

In addition to environmental values, Simon and Moltz [33] argue that political ideology and political party identification are significant moderators of public opinion about funding proposals in the areas of the natural environment, science, and alternative energy. In the area of climate change research, there are consistent findings that Democrats and more liberally-minded individuals are perceived to be more supportive of climate policies than Republicans and more conservatively-minded individuals [34–38]. Yet, there is a lack of investigation into the role of political ideology in shaping public opinion on energy security and alternative energy sources in the United States.

1.2. Knowledge

A review of the relevant literature demonstrates conflicting results about the role of policy relevant knowledge factors in influencing public opinion. Pierce et al. [39] found that more energy-informed citizens were more supportive of renewable energy policies. Hobman and Ashworth [31] discovered that a provision of additional information about a range of alternative energy technologies leads to greater public support for the use of said technologies. At the same time, Wolske et al. [27] contend that more information about carbon removal technologies may actually discourage public support, due to learning about new risks and potential impacts.

1.3. Socio-Demographic Characteristics

Steel et al. [32] show that younger and more educated respondents are more likely to support government policies related to clean energy technologies. Pierce and Steel [40] find that women and older individuals display a greater opposition towards alternative energy technologies. In regard to the public opinion on energy security, Knox-Hayes et al. [6] argue that women, less educated, and older individuals are more concerned over energy security. In this research, we investigate the following socio-demographic characteristics: age, gender, education, and income.

Our research objective is to understand how environmental value orientations, knowledge factors and socio-demographic characteristics are associated with concerns over energy security and public perceptions of new energy technologies in the U.S. context.

2. Materials and Methods

To address our research objectives, public opinion survey data were collected through household surveys conducted in California, Oregon, Washington, and Idaho in 2013. These states were selected because of their commitment to and investment in new clean energy technologies as part of their participation in the 2008 Pacific Coast Collaborative (PCC). The PCC is a regional approach to solving policy issues such as environmental protection and climate change, which has led the states to pursue aggressive renewable portfolio standards (RPS) and policies that encourage innovation in renewable energy technologies. In 2016, PCC states and the Canadian Province of British Columbia signed the 2016 Pacific Coast Climate Leadership Action Plan, which updated efforts at greenhouse gas emissions mitigation and adoption of community-scale renewable energy technologies. The state of Idaho is included as a control comparison. While it is also in the U.S. west and borders Oregon and Washington and is also heavily reliant on cheap energy from hydroelectric sources, it is more politically conservative and has not pursued state policies that promote the development and implementation of renewable energy technologies.

A mail survey with an additional link to an online option was sent to random samples of over 1400 households in each state. Even under the most strict sampling rules, assuming a 50/50 split in the population (i.e., 50% answer one way, while 50% answer the other way), to be 95% confident that an estimate from a sample survey is within +/− 3 percentage points of the true population value, a random sample of 1067 is needed for a population of 1 million and over [41]. Therefore, our sample size meets accepted standards of survey design. Samples were provided by a commercial research company that has exhaustive databases of households comprised of telephone directories, state departments of motor vehicle records, and other household information sources. Dillman's [41] Tailored Design Method was used in questionnaire design and implementation, which includes multiple reminder waves for non-responses and structured survey instruments and cover letters. A systematic sampling approach was applied within each household by asking those residents with the most recent birthday and over 18 years old to take the survey. Three waves of the mail questionnaires were distributed, followed by a final telephone reminder. Survey response rates vary only marginally across the four states, with the highest percentage in Oregon (51.5%), followed by 48.9% in Washington, 48.3% for California, and 46.6% for Idaho. Given the nature of the questions in the survey and the protections in place to protect individual respondent's identities, the Oregon State University Institutional Research Board determined that the research was "exempt" and therefore did not require full board review for ethical concerns.

In terms of survey response bias, we compared demographic data from the U.S. Census to survey data (Table 1). The Census data used is only for the section of the population that is 18 years and older as this aligns with the samples used. Survey respondents are slightly more affluent, older, and educated when compared to the Census data for each state. This finding is typical for survey research respondents [42]. The percentage of female and male respondents is almost identical to the Census data for all four states.

Table 1. Survey Response Bias.

California		
Demographic Variable	**Survey Sample**	**Census Estimates** [1]
Mean Age (Over 18)	47.7	47.1
Median Household Income	\$50,000–\$74,999 (Survey category 6)	\$60,883 (2006–2010 adjusted average)
Gender (Over 18)	Male 51.3%, Female 48.7%	Male 49.5%, Female 51.5%
Associates Degree or Higher (Over 18)	40.3%	36.7%
Idaho		
Demographic Variable	**Survey Sample**	**Census Estimates** [1]
Mean Age (Over 18)	52.6	48.0
Median Household Income	\$50,000–\$74,999 (Survey category 6)	\$46,890 (2006–2010 adjusted average)
Gender (Over 18)	Male 49.9%, Female 50.1%	Male 50%, Female 50%
Associates Degree or Higher (Over 18)	42.3%	39.1%
Oregon		
Demographic Variable	**Survey Sample**	**Census Estimates** [1]
Mean Age (Over 18)	55.3	49.5
Median Household Income	\$50,000–\$74,999 (Survey category 6)	\$49,260 (2006–2010 adjusted average)
Gender (Over 18)	48.7% Male, 51.3% Female	48.4% Male, 51.6% Female
Associates Degree or Higher (Over 18)	38.1%	35.0%
Washington:		
Demographic Variable	**Survey Sample**	**Census Estimates** [1]
Mean Age (Over 18)	50.3	48.5
Median Household Income	\$50,000–\$74,999 (Survey category 6)	\$57,224 (2006–2010 adjusted average)
Gender (Over 18)	48.3% Male, 51.7% Female	48.7% Male, 51.3% Female
Associates Degree or Higher (Over 18)	44.8%	38.8%

[1] Data obtained from the U.S. 2010 American Community Survey.

3. Results

Measures related to the concern over energy supply, being personally affected by energy shortage, support for government investments into research and development of alternative energies, perceptions of new energy technologies, political ideology, environmental beliefs, knowledge about energy, and socio-demographic characteristics were formed from survey responses. The survey questions used to create variables are provided in Appendix A. See Appendix B for descriptive statistics for all measures.

To assess how informed the public is about energy policy, we asked respondents to report their level of familiarity with renewable energy policy. Response categories were oriented on a four-point scale ranging from 1 = "Not informed" to 4 = "Very well informed" (mean = 2.12). To assess respondents' knowledge about energy, we asked three energy-specific questions: (1) what is the largest source of energy for electricity in your state?; (2) what economic sector uses the greatest share of electricity in your state?, and (3) what does it mean to be "off-grid"? Answers to these questions were formed into a Quiz index ranging from 0 = no correct answers to 3 = three correct answers (mean = 1.09).

Ideology was measured on a five-point scale from liberal to conservative (1 = "Very liberal" to 5 = "Very conservative"; mean = 3.03). Environmental values were measured using the New Ecological Paradigm (NEP) six-item scale. Answers ranged from 6 = low level of support for NEP to 30 = high level of support for NEP (mean = 21.02; see Appendix A).

Demographic variables included the gender of the respondent (male vs. female; 51% female), age in years (mean = 49), income on a 10-point scale (1 = "less than \$10,000" to 10 = "\$200,000 or more"; mean = 5.32) and formal education attainment on an 8-point scale (1 = "less than high school" to 8 = "postgraduate degree"; mean = 5.17).

Descriptive statistics for questions about public perceptions of energy scarcity and electricity shortage reveal within sample and across state variation (Table 2). The difference between states is not statistically significant for the question about national energy resources (Chi-square = 11.094, $p = 0.521$), but is statistically significant for the question about concern over personal energy scarcity (Chi-square = 33.092, $p = 0.001$). The majority of respondents (over 50%) in all states agree or strongly agree that the country does not have a sufficient supply of energy resources. The largest percent of respondents who agree with this statement live in California, while the largest percent of people who disagree live in Idaho. Regarding the concern about being personally affected by electricity shortages in the next 5 years, there is significant variation across states. Yet, similarly to the previous question, respondents from California and Oregon express a higher level of concern compared to respondents from Idaho and Washington. Additional Chi-square tests comparing state by state separately revealed that for concern about being personally affected by a shortage of electricity, Californian respondents were significantly more concerned in each state-by-state comparison. Perhaps this is not surprising given the brownouts and power outages Californians have experienced over the past decade [40]. In addition, Idaho respondents were significantly different from each of the states, with fewer respondents being concerned about possible future power shortages. This may be attributable to the abundant, dependable, and low cost hydroelectricity available to most Idaho residents [40].

Table 2. Public perceptions of energy scarcity; variation across states.

Question: How much do you agree or disagree with the following statements concerning energy policy?				
"I am concerned that our country doesn't have enough energy resources."				
	California	**Idaho**	**Oregon**	**Washington**
	Percent	Percent	Percent	Percent
Strongly Disagree	11.9	12.1	9.9	10.1
Disagree	16.4	19.1	16.3	18.0
Neutral	12.9	12.8	14.5	16.3
Agree	27.3	26.7	29.6	27.0
Strongly Agree	31.4	29.2	29.7	28.6
N =	688	685	754	711
Chi-square =	11.094, $p = 0.521$			
"I am concerned about being personally affected by shortage of electricity in the next five years."				
Strongly Disagree	11.0	8.6	11.8	11.1
Disagree	20.7	26.6	21.1	27.7
Neutral	24.8	28.5	29.4	27.9
Agree	25.7	23.6	24.5	22.0
Strongly Agree	17.8	12.7	13.2	11.3
N =	690	687	755	714
Chi-square =	33.092, $p = 0.001$			

Evaluating the descriptive statistics of public perceptions of new energy technologies, we observe that the responses are skewed toward agree and strongly agree answers for both statements: (1) that not enough money is being spent on research and development of alternative fuels and (2) that new technologies will make it possible to have enough electricity for all in the future (Table 3). Similar to the findings about perceptions of energy scarcity, a larger percentage of respondents from California and Oregon expressed concern over the level of funding for research and development. Also, a larger proportion of respondents from California and Oregon believed in the future potential of new energy technologies, compared to respondents from Idaho and Washington. In both cases the difference between states is statistically significant (Chi-square = 23.466, $p = 0.024$ and Chi-square = 21.925, $p = 0.038$, respectively).

Table 3. Public perceptions of new energy technologies; variation across states.

Question: How much do you agree or disagree with the following statements concerning energy policy?				
"Not enough money is being spent on research and development of alternative fuels."				
	California	**Idaho**	**Oregon**	**Washington**
	Percent	Percent	Percent	Percent
Strongly Disagree	3.8	6.1	5.9	5.1
Disagree	11.1	14.9	10.1	12.9
Neutral	22.7	21.3	20.2	25.5
Agree	29.0	29.2	30.7	28.1
Strongly Agree	33.5	28.5	33.1	28.4
N =	687	685	752	711
Chi-square =	23.466, $p = 0.024$			
"New technologies will make it possible to have enough electricity for all of us in the future."				
Strongly Disagree	1.0	2.8	2.5	2.7
Disagree	4.7	9.5	7.3	6.9
Neutral	18.3	17.2	19.7	18.7
Agree	39.4	37.1	35.8	38.7
Strongly Agree	36.6	33.5	34.7	33.1
N =	688	687	755	713
Chi-square =	21.925, $p = 0.038$			

As with the analyses presented in Table 2, additional Chi-square tests were conducted for state-to-state comparisons. Concerning the statement that not enough money is being spent on research and development, California and Oregon respondents were not significantly different in their responses, and the same can be said with Idaho and Washington respondents. However, the Chi-square analyses showed that California and Oregon respondents were significantly different from Idaho and Washington survey participants in their level of agreement and disagreement with the statement. California and Oregon respondents were slightly less like to disagree with the statement and more likely to agree.

For the final statement in Table 2, concerning new technologies contributing to electricity for all in the future, the additional Chi-square results show that California respondents were significantly different from the other three states in their agreement with the statement. While over 70 percent of respondents in each state agreed or strongly agreed with the statement, Californians were significantly less likely to disagree with the statement and more likely to agree with the statement when compared to each other state separately.

Due to skewed distribution of dependent variables, measures were recoded into binary variables (1 = agree, 0 = else) and a logistic regression analysis was performed to estimate the relationships between dependent and explanatory variables. Table 4 highlights results of the logistic regression output for two dependent variables: concern over energy scarcity and concern over personal energy shortage. Among socio-demographic factors, the findings indicate that being female and having a higher level of formal education is significantly associated with a lower level of concern over energy scarcity, while higher income is significantly associated with a lower level of personal concern over energy shortage. For the knowledge variables, respondents who are more familiar with renewable energy policy are less likely to be concerned over energy scarcity, while those with a better performance on an energy quiz have lower levels of concern over personal energy shortage. Among value and ideology factors, a higher score on the New Ecological Paradigm scale is associated with greater concerns about U.S. energy security, as well as personal energy security. Finally, being more politically conservative has shown to be associated with greater concern over personal energy security.

Table 4. Logistic regression estimates for energy security beliefs.

	Concern That the Country Does Not Have Enough Energy Resources [a]	Concern over Being Personally Affected by Energy Shortage [b]
	Coefficient (S.E.)	Coefficient (S.E.)
Age	−0.004 (0.003)	−0.003 (0.003)
Gender	−0.228 ** (0.086)	−0.331 *** (0.087)
Education	−0.090 ** (0.035)	−0.123 *** (0.038)
Income	−0.011 (0.020)	−0.053 ** (0.020)
Familiar	−0.287 *** (0.055)	0.079 (0.055)
Quiz	−0.054 (0.058)	−0.217 *** (0.058)
NEP	0.060 *** (0.009)	0.042 *** (0.009)
Ideology	−0.089 (0.048)	0.223 *** (0.048)
N =	2641	2648
Chi−square =	166.438 ***	115.498 ***
Percent correctly predicted =	63.3%	63.7%
Nagelkerke R2	0.082	0.058
	$^{**}p \leq 0.01$; $^{***}p < 0.001$	

[a] 1 = Agree that country does not have enough energy resources, 0 = else. [b] 1 = Agree will be personally affected by energy shortage, 0 = else. NEP = New Ecological Paradigm.

Table 5 presents results of the logistic regression for the second set of dependent variables on energy technology beliefs. Here, we discover diverging results regarding the influence of gender. Females express greater concern that not enough money is spent on research and development of technologies. At the same time, they are less likely to think that technologies will provide energy for all in the future. Respondents with more advanced formal education are less likely to believe in the impact of technology on future energy supply, while higher income level is associated with lower level of concern that not enough resources are being spent on research and development. Concerning the impact of environmental values, those respondents with higher NEP scores are more likely to agree that not enough money is being spent on research and development. Finally, being politically conservative is associated with lower levels of concerns over the shortage of funding for research and development of new energy technologies and lower levels of perception that technologies will supply energy for all in the future.

Table 5. Logistic regression estimates for energy technology beliefs.

	Not Enough Money Spent on Research and Development [a]	New Technologies Make It Possible to Have Energy for All in the Future [b]
	Coefficient (S.E.)	Coefficient (S.E.)
Age	0.000 (0.003)	0.001 (0.003)
Gender	0.253 ** (0.094)	−0.202 * (0.093)
Education	0.008 (0.037)	−182 *** (0.039)
Income	−0.045 * (0.022)	0.031 (0.021)
Familiar	−0.042 (0.059)	0.004 (0.058)
Quiz	0.113 (0.062)	−0.115 (0.062)
NEP	0.115 *** (0.010)	0.008 (0.010)
Ideology	−0.520 *** (0.053)	−0.176 *** (0.052)
N =	2640	2646
Chi-square =	472.019 ***	55.279 ***
Percent correctly predicted =	67.9%	71.9%
Nagelkerke R2	0.222	0.030
	* $p \leq 0.05$; ** $p \leq 0.01$; *** $p \leq 0.001$	

[a] 1 = Agree that not enough money spent on research and development of alternative fuels, 0 = else. [b] 1 = Agree new technologies will make it possible to have electricity for all in the future, 0 = else.

4. Discussion

4.1. Environmental Value Orientations and Ideology

Reflecting results of previous studies that show a connection between environmental value orientations and pro-environmental behaviors, such as displaying positive attitudes for new alternative energy sources [31] and government investments in alternative energy [32], this study finds that pro-environmental values are associated with public perceptions that not enough resources are being devoted to research and development of new energy technologies and greater concerns about the U.S. energy security, as well as personal energy security. At the same time, these respondents do not seem to support the idea that new technologies can ensure energy supply for all in the future. It is possible that respondents with higher biocentric scores on the NEP scale are concerned about the potential negative impacts of new technologies on the environment [24–27], and thus, the extent to which technologies should serve as a solution to energy problems in the future. Building on the research by Simon and Moltz [33], who contend that political ideology is a strong predictor of public opinion about government spending in areas of environment and technologies, we demonstrate that conservatives are less concerned about the insufficiency of government funding towards research and development of new energy technologies, and are also less likely to believe that alternative energy technologies are capable of being an adequate energy resource in the future. Government investment in new energy technologies implies a number of politically sensitive issues concerning the role of government involvement in the energy market and growth of renewable energy market share. Our findings suggest that conservative leaning respondents are reluctant to provide government support for new energy technologies [40]. At the same time, conservatives also displayed higher concern about experiencing personal energy shortages.

4.2. Knowledge

Similar to prior studies on the connection between knowledge and public opinion about new technologies [27,31,40], we found that greater familiarity with renewable energy policy is associated with lower concerns over the country's energy scarcity. It is possible that respondents who are more familiar with renewable energy policy have a better understanding of energy policy in general and, therefore, are confident in the ability of the market and the government to ensure a reliable energy supply in the future, regardless of the type of energy technologies employed to accomplish that. As we show, trust in new energy technology's ability to provide energy supply for all in the future is not associated with renewable energy policy familiarity. In regard to the energy knowledge quiz, respondents who scored higher on the quiz, have fewer concerns about being personally affected by the electricity shortage in the next 5 years. It is worth mentioning that questions on the quiz were state-specific. Therefore, our findings showcase an idea that being informed about local energy issues is associated with lower levels of concern about being personally affected by shortages of energy supply. Interestingly, neither familiarity with renewable energy policy nor energy knowledge variables are associated with perception of the level of government funding of new technologies or the power of new technologies to ensure a sustainable supply of energy in the future. This discovery suggests a diversion from previous research findings on the connection between knowledge and public opinion about new technologies [27,31,40]. We establish that familiarity with general energy issues and renewable energy policy is not necessarily associated with public perceptions on government investments into new energy technologies or on the technical capabilities of those technologies.

4.3. Socio-Demographic Characteristics

Contrary to findings by Knox-Hayes et al. [6], we discover that women are less concerned about energy security issues when compared to men. This is an interesting finding, because a number of studies in sociology and psychology demonstrate systematic differences between men and women in attitudes toward risk, arguing that on average women tend to be more risk averse [43]. Thus, in our work, we would expect women to be more concerned about the energy security issue than men. However, as Eckel and Grossman [43] contend, when looking at gender attitudes toward risk, it is important to account for other demographic factors such as knowledge, wealth, marital status and others. It is possible that in our study women are less concerned about energy security issues because our sample is slightly more affluent and with higher level of education than the population. At the same time, women are also more likely to perceive a shortage of government funding towards research and development of new technologies. Attesting to the connection among the demographic factors, we find that more educated respondents are less likely to be concerned over energy security in the future, personally and for the nation as a whole. It is possible that respondents with a higher level of education enjoy higher incomes, and therefore, a greater sense of personal security over any future event. To support this statement, we show that those with higher incomes are less concerned about being personally impacted by electricity shortage in the next 5 years. Furthermore, respondents with higher levels of formal education are less likely to believe in the power of new technologies to support a reliable supply of energy in the future. It is possible that the more educated public accepts a more cautious view about the successful and rapid integration of new technologies into the market. As we observe, respondents leaning toward conservative political views also take on a more reserved stance about the feasibility of new energy technologies securing a sustainable supply of energy in the future. Finally, age did not play a role across any of the analyzed opinions. This is an interesting finding, as we may expect that the respondents belonging to the generation that lived through the oil crisis of the 1970s, a period infamous for oil shortages and high energy prices [44], would be more concerned about energy shortages in the future. At the same time, we may also assume that a younger generation would be leaning toward higher trust of new energy technologies.

Author Contributions: Contributions of the individual authors in this work are the following: conceptualization, B.S.S. and C.A.S.; methodology, B.S.S.; formal analysis, B.S.S. and A.B.; resources, B.S.S. and C.A.S.; data curation, B.S.S.; writing—original draft preparation, A.B.; writing—review and editing, A.B., B.S.S., and C.A.S.; visualization, B.S.S. and A.B.; supervision, B.S.S.; project administration, B.S.S.; funding acquisition, B.S.S. All authors have read and agreed to the published version of the manuscript.

Funding: This research was funded by the Oregon Policy Analysis Laboratory, School of Public Policy, Oregon State University, and the United State Department of Agriculture grant "Climate Change Adaptation, Sustainable Energy Development and Comparative Agricultural and Rural Policy," National Institute of Food and Agriculture (NIFA) Higher Education Challenge (HEC) Grants Program.

Acknowledgments: We would like to thank former master students in the School of Public Policy at Oregon State University: Mariana Amorim, Courtney Flathers, Andrew Spaeth, Lyndsay Trant, and Iaroslav Vugniavyi for their contributions to the data collection.

Conflicts of Interest: The authors have declared that no competing interests exist.

Appendix A

Dependent Variables

How much do you agree or disagree with the following statements concerning energy policy?

	Strongly Disagree	Somewhat Disagree	Neutral	Somewhat Agree	Strongly Agree
I am concerned that our country doesn't have enough energy resources.	1	2	3	4	5
I am concerned about being personally affected by shortage of electricity in the next five years.	1	2	3	4	5
Not enough money is being spent on research and development of alternative fuels.	1	2	3	4	5
New technologies will make it possible to have enough electricity for all of us in the future.	1	2	3	4	5

Sociodemographic Variables

We now have a few concluding questions to check if our survey is representative of all types of people. Please remember that all answers are completely confidential to the extent permitted by law.

What is your current age in years ___________?
Please indicate your gender? 1. Female 2. Male
What level of education have you completed?

1. Grade School	5. Some college
2. Middle or junior high school	6. College graduate
3. High school	7. Graduate school
4. Vocational school	8. Other ______________________?

Which category best describes your household income (before taxes) in 2014?

1. Less than $10,000	6. $50,000–$74,999
2. $10,000–$14,999	7. $75,000–$99,999
3. $15,000–$24,999	8. $100,000–$149,999
4. $25,000–$34,999	9. $150,000–$199,999
5. $35,000–$49,999	10. $200,000 or more

Knowledge Questions

Familiarity:

In general, how well informed would you consider yourself to be concerning renewable energy policy issues in (state)—such as wind, solar, wave, and biomass energy?

1. Not informed
2. Somewhat informed
3. Informed
4. Very well informed

Energy Quiz:

Here are a few specific questions about energy. Many people don't know the answers to these questions, so if there are some you don't know just leave them blank and continue.	
	a. The largest source of energy for electricity in your state is:
1.	Coal
2.	Hydroelectric
3.	Natural Gas
4.	Nuclear
	b. Most electricity in your state is used by the:
1.	Residential Sector (e.g., households)
2.	Commercial Sector (e.g., retail stores)
3.	Industrial Sector (e.g., factories and mills)
4.	Transportation Sector
	c. Being "off-grid" means:
1.	Producing one's own electricity
2.	Getting electricity from another state
3.	Having no electricity
4.	Being energy efficient

The Quiz variable is an additive index of correct answers. Correct answers are: (a) Idaho, Oregon and Washington–hydroelectric; California–natural gas; (b) California, Oregon and Washington–transportation; Idaho–industrial; (c) Producing one's own electricity.

New Ecological Paradigm Index					
Listed below are statements about the relationship between humans and the environment. For each, please indicate your level of agreement.					
	Strongly Disagree	**Mildly Disagree**	**Neutral**	**Mildly Agree**	**Strongly Agree**
The balance of nature is very delicate and easily upset by human activities.	1	2	3	4	5
Humans have the right to modify the natural environment to suit their needs.	1	2	3	4	5
We are approaching the limit of people the earth can support.	1	2	3	4	5
The so-called "ecological crisis" facing humankind has been greatly exaggerated.	1	2	3	4	5
Plants and animals have as much right as humans to exist.	1	2	3	4	5
Humans were meant to rule over the rest of nature	1	2	3	4	5
Statements 2, 4 and 6 above were recoded to: 5 = biocentric response and 1 = anthropocentric response. The items were then used in an additive index that ranges from 6 to 30. Chronbach's alpha is 0.759.					

Political Ideology On domestic policy issues, would you consider yourself to be?				
1. Very Liberal	2. Liberal	3. Moderate	4. Conservative	5. Very Conservative

Appendix B

Variable	Questions/Categories	Descriptive Statistics
	Dependent Variables	
Concern over energy supply	*"I am concerned that our country doesn't have enough energy resources."* Categories ranging from 1 = "Strongly disagree" to 5 = "Strongly Agree"	mean = 3.477 std.dev. = 1.36 N = 2838
Being personally affected by energy shortage	*"I am concerned about being personally affected by shortage of electricity in the next five years."* Categories ranging from 1 = "Strongly disagree" to 5 = "Strongly Agree"	mean = 3.06 std.dev. = 1.2 N = 2846
Research and development	*"Not enough money is being spent on research and development of alternative fuels."* Categories ranging from 1 = "Strongly disagree" to 5 = "Strongly Agree"	mean = 3.68 std.dev. = 1.18 N = 2835
New technologies will ensure future energy supply	*"New technologies will make it possible to have enough electricity for all of us in the future."* Categories ranging from 1 = "Strongly disagree" to 5 = "Strongly Agree"	mean = 3.95 std.dev. = 1.01 N = 2843

Independent variables		
Age	Age in years (range = 18 to 98)	mean = 49.3 s.d. = 16.10 N = 2845
Gender	1 = female, 0 = male	mean = 0.51 N = 2840
Education	Formal educational attainment (1 = less than high school to 8 = postgraduate degree)	mean = 5.17 s.d. = 1.25 N = 2811
Income	Household income before taxes in 2017. (1 = less than $10,000 to 10 = $200,000 or more)	mean = 5.32 s.d. = 2.15 N = 2727
Informed about energy policy	Level of self-assessed familiarity with renewable energy policy. (1 = not informed to 4 = very well informed)	mean = 2.12 s.d.= 0.77 N = 2848
Quiz	Energy quiz score. (0 = no correct answers to 3 = three correct answers)	mean = 1.09 s.d. = 0.74 N = 2848
Ideology	Subjective political ideology (1 = Very liberal to 5 = Very conservative)	mean = 3.03 s.d. = 0.99 N = 2829
NEP	New Ecological Paradigm (6 = low level of support to 30 high level of support)	mean = 21.02 s.d. = 5.29 N = 2835

References

1. Ansolabehere, S.; Konisky, D.M. *Cheap and Clean: How Americans Think about Energy in the Age of Global Warming*; MIT Press: Cambridge, MA, USA, 2014.
2. Peterson, T.R.; Stephens, J.C.; Wilson, E.J. Public Perception of and Engagement with Emerging Low-Carbon Energy Technologies: A Literature Review. *MRS Energy Sustain.* **2015**, 2. [CrossRef]
3. Besley, J.C. The State of Public Opinion Research on Attitudes and Understanding of Science and Technology. *Bull. Sci. Technol. Soc.* **2013**, *33*, 12–20. [CrossRef]
4. Ekins, P. Step Changes for Decarbonizing the Energy System: Research Needs for Renewables, Energy Efficiency and Nuclear Power. *Energy Policy* **2004**, *32*, 1891–1904. [CrossRef]
5. Sovacool, B.K. Differing Cultures of Energy Security: An International Comparison of Public Perceptions. *Renew. Sustain. Energy Rev.* **2016**, *55*, 811–822. [CrossRef]
6. Knox-Hayes, J.; Brown, M.A.; Sovacool, B.K.; Wang, Y. Understanding Attitudes toward Energy Security: Results of a Cross-National Survey. *Glob. Environ. Chang.* **2013**, *23*, 609–622. [CrossRef]
7. DeCicco, J.; Yan, T.; Keusch, F.; Muñoz, D.H.; Neidert, L.U.S. Consumer Attitudes and Expectations about Energy. *Energy Policy* **2015**, *86*, 749–758. [CrossRef]
8. Crainz, M.; Curto, D.; Franzitta, V.; Longo, S.; Montana, F.; Musca, R.; Riva Sanseverino, E.; Telaretti, E. Flexibility Services to Minimize the Electricity Production from Fossil Fuels. A Case Study in a Mediterranean Small Island. *Energies* **2019**, *12*, 3492. [CrossRef]
9. Curto, D.; Franzitta, V.; Viola, A.; Cirrincione, M.; Mohammadi, A.; Kumar, A. A Renewable Energy Mix to Supply Small Islands. A Comparative Study Applied to Balearic Islands and Fiji. *J. Clean. Prod.* **2019**, *241*, 118356. [CrossRef]
10. Scheer, D.; Konrad, W.; Scheel, O. Public Evaluation of Electricity Technologies and Future Low-Carbon Portfolios in Germany and the USA. *Energy Sustain. Soc.* **2013**, *3*, 8. [CrossRef]
11. Sütterlin, B.; Siegrist, M. Public Acceptance of Renewable Energy Technologies from an Abstract versus Concrete Perspective and the Positive Imagery of Solar Power. *Energy Policy* **2017**, *106*, 356–366. [CrossRef]

12. Reiner, D.M.; Curry, T.E.; de Figueiredo, M.A.; Herzog, H.J.; Ansolabehere, S.D.; Itaoka, K.; Johnsson, F.; Odenberger, M. American Exceptionalism? Similarities and Differences in National Attitudes Toward Energy Policy and Global Warming. *Environ. Sci. Technol.* **2006**, *40*, 2093–2098. [CrossRef] [PubMed]
13. Bigerna, S.; Bollino, C.A.; Micheli, S.; Polinori, P. A New Unified Approach to Evaluate Economic Acceptance towards Main Green Technologies Using the Meta-Analysis. *J. Clean. Prod.* **2017**, *167*, 1251–1262. [CrossRef]
14. Lachapelle, E.; Borick, C.P.; Rabe, B. Public Attitudes toward Climate Science and Climate Policy in Federal Systems: Canada and the United States Compared1. *Rev. Policy Res.* **2012**, *29*, 334–357. [CrossRef]
15. Wennersten, R.; Sun, Q.; Li, H. The Future Potential for Carbon Capture and Storage in Climate Change Mitigation—An Overview from Perspectives of Technology, Economy and Risk. *J. Clean. Prod.* **2015**, *103*, 724–736. [CrossRef]
16. Rand, J.; Hoen, B. Thirty Years of North American Wind Energy Acceptance Research: What Have We Learned? *Energy Res. Soc. Sci.* **2017**, *29*, 135–148. [CrossRef]
17. Giordono, L.S.; Boudet, H.S.; Karmazina, A.; Taylor, C.L.; Steel, B.S. Opposition "Overblown"? Community Response to Wind Energy Siting in the Western United States. *Energy Res. Soc. Sci.* **2018**, *43*, 119–131. [CrossRef]
18. Eaton, W.M. What's the Problem? How 'Industrial Culture' Shapes Community Responses to Proposed Bioenergy Development in Northern Michigan, USA. *J. Rural Stud.* **2016**, *45*, 76–87. [CrossRef]
19. Klick, H.; Smith, E.R.A.N. Public Understanding of and Support for Wind Power in the United States. *Renew. Energy* **2010**, *35*, 1585–1591. [CrossRef]
20. Wiersma, B.; Devine-Wright, P. Public Engagement with Offshore Renewable Energy: A Critical Review. *Wiley Interdiscip. Rev. Clim. Chang.* **2014**, *5*, 493–507. [CrossRef]
21. Raimi, K.T.; Carrico, A.R. Understanding and Beliefs about Smart Energy Technology. *Energy Res. Soc. Sci.* **2016**, *12*, 68–74. [CrossRef]
22. Spence, A.; Demski, C.; Butler, C.; Parkhill, K.; Pidgeon, N. Public Perceptions of Demand-Side Management and a Smarter Energy Future. *Nat. Clim Chang.* **2015**, *5*, 550–554. [CrossRef]
23. Krause, R.M.; Carley, S.R.; Lane, B.W.; Graham, J.D. Perception and Reality: Public Knowledge of Plug-in Electric Vehicles in 21 U.S. Cities. *Energy Policy* **2013**, *63*, 433–440. [CrossRef]
24. Bassarak, C.; Pfister, H.-R.; Böhm, G. Dispute and Morality in the Perception of Societal Risks: Extending the Psychometric Model. *J. Risk Res.* **2017**, *20*, 299–325. [CrossRef]
25. Corner, A.; Parkhill, K.; Pidgeon, N.; Vaughan, N.E. Messing with Nature? Exploring Public Perceptions of Geoengineering in the UK. *Glob. Environ. Chang.* **2013**, *23*, 938–947. [CrossRef]
26. Vandermoere, F.; Blanchemanche, S.; Bieberstein, A.; Marette, S.; Roosen, J. The Morality of Attitudes toward Nanotechnology: About God, Techno-Scientific Progress, and Interfering with Nature. *J. Nanopart. Res.* **2010**, *12*, 373–381. [CrossRef]
27. Wolske, K.S.; Raimi, K.T.; Campbell-Arvai, V.; Hart, P.S. Public Support for Carbon Dioxide Removal Strategies: The Role of Tampering with Nature Perceptions. *Clim. Chang.* **2019**, *152*, 345–361. [CrossRef]
28. Dunlap, R.E.; Liere, K.D.V.; Mertig, A.G.; Jones, R.E. New Trends in Measuring Environmental Attitudes: Measuring Endorsement of the New Ecological Paradigm: A Revised NEP Scale. *J. Soc. Issues* **2000**, *56*, 425–442. [CrossRef]
29. Stern, P.C.; Dietz, T.; Abel, T.; Guagnano, G.A.; Kalof, L. A Value-Belief-Norm Theory of Support for Social Movements: The Case of Environmentalism. *Hum. Ecol. Rev.* **1999**, *6*, 81–97.
30. Jaeger, C.C.; Webler, T.; Rosa, E.A.; Renn, O.; Webler, T.; Rosa, E.A.; Renn, O. *Risk, Uncertainty and Rational Action*; Routledge: Abingdon, UK, 2013. [CrossRef]
31. Hobman, E.V.; Ashworth, P. Public Support for Energy Sources and Related Technologies: The Impact of Simple Information Provision. *Energy Policy* **2013**, *63*, 862–869. [CrossRef]
32. Steel, B.S.; Pierce, J.C.; Warner, R.L.; Lovrich, N.P. Environmental Value Considerations in Public Attitudes About Alternative Energy Development in Oregon and Washington. *Environ. Manag.* **2015**, *55*, 634–645. [CrossRef]
33. Simon, C.A.; Moltz, M.C. Going Native? Examining Nativity and Public Opinion of Environment, Alternative Energy, and Science Policy Expenditures in the United States. *Energy Res. Soc. Sci.* **2018**, *46*, 296–302. [CrossRef]
34. Leiserowitz, A. Climate Change Risk Perception and Policy Preferences: The Role of Affect, Imagery, and Values. *Clim. Chang.* **2006**, *77*, 45–72. [CrossRef]

35. McCright, A.M.; Dunlap, R.E.; Xiao, C. Perceived Scientific Agreement and Support for Government Action on Climate Change in the USA. *Clim. Chang.* **2013**, *119*, 511–518. [CrossRef]
36. Mildenberger, M.; Marlon, J.R.; Howe, P.D.; Leiserowitz, A. The Spatial Distribution of Republican and Democratic Climate Opinions at State and Local Scales. *Clim. Chang.* **2017**, *145*, 539–548. [CrossRef]
37. Druckman, J.N.; McGrath, M.C. The Evidence for Motivated Reasoning in Climate Change Preference Formation. *Nat. Clim Chang.* **2019**, *9*, 111–119. [CrossRef]
38. Dunlap, R.E.; McCright, A.M. A Widening Gap: Republican and Democratic Views on Climate Change. *Environ. Sci. Policy Sustain. Dev.* **2008**, *50*, 26–35. [CrossRef]
39. Pierce, J.C.; Steel, B.S.; Warner, R.L. Knowledge, Culture, and Public Support for Renewable-Energy Policy. *Comp. Technol. Transf. Soc.* **2009**, *7*, 270–286. [CrossRef]
40. Pierce, J.C.; Steel, B.S. *Prospects for Alternative Energy Development in the U.S. West: Tilting at Windmills*; Springer: Berlin/Heidelberg, Germany, 2017.
41. Dillman, D.A.; Smyth, J.D.; Christian, L.M. *Internet, Phone, Mail, and Mixed-Mode Surveys: The Tailored Design Method*; John Wiley & Sons: Hoboken, NJ, USA, 2014.
42. American Association for Public Opinion Research. Final Dispositions of Case Codes and Outcome Rates for Surveys. 2011. Available online: http://www.aapor.org/AM/Template.cfm?Section=Standard_Definitions2&Template=/CM/ContentDisplay.cfm&ContentID=3156 (accessed on 15 April 2019).
43. Eckel, C.C.; Grossman, P.J. Chapter 113 Men, Women and Risk Aversion: Experimental Evidence. In *Handbook of Experimental Economics Results*; Plott, C.R., Smith, V.L., Eds.; Elsevier: Amsterdam, The Netherlands, 2008; Volume 1, pp. 1061–1073. [CrossRef]
44. Bolsen, T.; Cook, F.L. The Polls—TrendsPublic Opinion on Energy Policy: 1974–2006. *Public Opin. Q.* **2008**, *72*, 364–388. [CrossRef]

Article

A Conceptual Framework to Understand Households' Energy Consumption

Véronique Vasseur [1,*], Anne-Francoise Marique [2] and Vladimir Udalov [3]

1 International Centre for Integrated Assessment and Sustainable Development, University Maastricht, 6200 Maastricht, The Netherlands
2 Local Environnent: Management & Analysis, ArGenCo Department, University of Liège, 4000 Liège, Belgium; afmarique@ulg.ac.be
3 European Institute for International Economic Relations, Rainer-Gruenter-Str. 21, 42119 Wuppertal, Germany; Vladimir_udalov@yahoo.de
* Correspondence: veronique.vasseur@maastrichtuniversity.nl; Tel.: +31-(0)-43-388-3223

Received: 5 October 2019; Accepted: 4 November 2019; Published: 7 November 2019

Abstract: Households' energy consumption has received a lot of attention in debates on urban sustainability and housing policy due to its possible consequences for climate change. In Europe, the residential sector accounts for roughly one third of the energy consumption and is responsible for 16% of total CO_2 emissions. Households have been progressively highlighted as the main actor that can play a substantial in the reduction of this energy use. Their behavior is a complex and hard to change process that combines numerous determinants. These determinants have already been extensively studied in the literature from a variety of thematic domains (psychology, sociology, economics, and engineering), however, each approach is limited by its own assumptions and often omit important energy behavioral components. Therefore, energy behavior studies require an integration of disciplines through interdisciplinary approaches. Based on that knowledge, this paper introduces a conceptual framework to capture and understand households' energy consumption. The paper aims at connecting objective (physical and technical) with subjective (human) aspects related to energy use of households. This combination provide the answers to the 'what', the 'how' and most importantly the 'why' questions about people's behavior regarding energy use. It allows clarifying the numerous internal and external factors that act as key determinants, as well as the need to take into account their interactions. By doing so, we conclude the paper by discussing the value of the conceptual framework along with valuable insights for researchers, practitioners and policymakers.

Keywords: households; energy consumption; pro-environmental behavior; conceptual framework

1. Introduction

Households' energy consumption has received a lot of attention in debates on urban sustainability and housing policy due to its possible consequences for climate change [1–3]. The residential sector is responsible for 17% of global CO_2 emissions in the world and constitutes the third-largest major energy consumer worldwide [3]. According to Brounen et al. [1], about 20% of total global energy demand originates from the requirements to heat, cool, and light residential dwellings. In Europe, the residential sector stands for roughly 30% of the energy consumption and is responsible for 16% of total CO_2 emissions. According to the Environmental Investigation Agency (EIA) [2], households in Europe accounted for 21% of the world's total residential energy consumption in 2012. Space heating is responsible for the most important part of energy used by households. In accordance with recent literature highlighting the strong relationships between building energy consumption, location, transportation and urban form [4–7], individual mobility is considered in this paper as part of the energy uses at the household level. Transportation indeed represents a significate part of households' energy consumption [8,9]. Last but not least, transportation in the sole sector, at the European level, in which energy consumption and related emissions of greenhouse gases is increasing.

Reducing domestic energy uses is necessary, especially to achieve the international and national commitments to significantly reduce carbon emissions. By 2050, the European Union should cut greenhouse gas (GHG) emissions to 80% below 1990 levels. The milestones to be achieved are 40% cuts by 2030 and 60% by 2040 [10]. Still according to this low-carbon economy roadmap [10], emissions from the building sector (houses and offices) could be cut by around 90% in 2050 by improving drastically three strategies: passive housing technologies for new building; refurbishing old buildings and substituting electricity and renewables for fossil fuels in heating, cooling and cooking. In this search for more energy efficiency in the domestic sector, three main strategies have been the focus on extensive review in the current literature and are namely summarized within the "Trias Energetica concept" developed by TU Delft [11], consisting in three consecutive steps:

(1) Reducing demand for energy by avoiding waste and implementing energy-saving measures
(2) Using sustainable sources of energy instead of finite fossils fuels (renewable energy)
(3) Producing and using fossil energy as efficient as possible

This framework, as well as recent research [5,12,13] put the focus on the crucial need to reduce energy demand as the first and most efficient way toward a sustainable future. As far as regulations and policies are concerned, there are numerous local, national and also international regulations and policies aiming to reduce energy demand by strict technical requirements. For buildings characteristics, the European directive on the energy performance of buildings came into force in 2002, and was progressively strengthen to impose, by 2018 for public buildings and by 2020 for all new buildings to be nearly zero energy buildings. Retrofitting the existing building stock has also been highlighted as the main target to achieve [10], especially in Europe where the renewal rate of buildings is low [14–17]. Regulations on maximum CO_2 emissions for private vehicles are also periodically strengthen whereas initiatives focused on changes in consumption patterns, and the use of energy in a greener way remain more limited.

In energy efficiency research, households have been progressively highlighted as the main actor that can play a substantial in the reduction of this energy use [18–20]. Households' energy consumption is a complex and hard to change process that combines numerous determinants. It is made up by different characteristics of the building and the neighborhood in which the household live, by the energy-using appliances and heating/cooling systems, but more importantly by a variety of internal and external factors, such as households' beliefs, values and attitudes, other people's behaviors, and various economic incentives.

For example, Jones et al. [21], based on Wei et al. [22] and a review of the literature summarized key determinants (here for space heating) into four main categories, as follows:

(1) Environmental factors: indoor and outdoor climate, wind pressure, etc.
(2) Building and system related factors: dwelling type, dwelling age, insulation level, type of heating system, fuel, control, etc.
(3) Occupant related factors: age, gender, education level, socio-economic classification, household size, etc.
(4) Others factors: occupancy, heating prices, awareness of energy use, and attitudes about energy use.

Each determinant considered alone, or some combinations, of determinants within the same category, has already been extensively studied in the literature and research on household energy consumption has mainly focused on the economic and technological aspects of this issue, while most of the policy action has aimed at reducing information barriers and providing financial incentives (see the literature overview for overview of the key literature that identify the factors affecting housholds' energy consumption). In this perspective dominated by neoclassical economics, a growing body of research in behavioral sciences and sociology showing that household energy consumption is far more complex than the assumptions made in cost-benefit analyses has largely been overlooked. Actually, it is formed by a combination of factors, not only individual factors but also contextual factors are of importance. Due to this complexity, household energy consumption is often studied using a more fragmented and disciplinary studies from a variety of thematic domains such as psychology, sociology, economics and engineering. While technological approaches focus on quantifying energy consumption as a support for decision-making, approaches in the social sciences focus on understanding and explaining actual energy behavior. Nonetheless, each approach is constrained by its own assumptions and it often omit important energy behavioral components. Therefore, energy behavior studies require an integration of different disciplines by using an interdisciplinary approach.

In this context, the aim of this paper is to introduce a new conceptual framework to capture and understand the households' energy consumption. The paper aims at connecting objective (physical and technical) with subjective (human) aspects related to energy use by households. This combination aims at providing the answers to the 'what', the 'how' and most importantly the 'why' questions about people's behavior regarding energy use. In order to underhand how households' energy consumption work, Section 2 firstly provides the methodology followed by review of exiting behavioral change theories analyzing and identifying strengths and weaknesses of the models (Section 3). Such analysis combines technical and behavioral determinants of energy consumption as well as environmental influence constituting a set of aspects which leads to develop a differentiation of the main aspects of households' energy consumption. Then, Section 4 proposes a new comprehensive conceptual framework concerning determinants of the external and internal context. Finally, Section 5 summarizes our main findings and highlights new insights and perspective for future research in households' behavior and energy efficiency.

2. Methodology

The first part of this research is a literature review in order to define more clearly what is to be examined, with the intention of having a sufficient outline for determining what data to collect and how to analyse the data in practice [23]. The literature review consists of 4 steps: (1) selection of papers; (2) preliminary analysis; (3) detailed analysis; and (4) framework development.

Step 1: Selection of papers

The literature were searched on Scopus and Web of Science online databases due to their ability to allow fast and customized searches. The basic terms for the review were identified as "energy efficiency" and "behavior", the first search on the database was performed using the "energy eff*" which included both "energy efficiency" and "energy efficient". Next, the search was limited to journal articles in English only. A further filtering based on title reviewing was carried out and we determined the articles relevant enough to be included in the analysis. The criteria used for the inclusion of the articles were the following:

- Studies where the energy efficiency concept is the main topic
- Publications that are focused on households' energy consumption / households' behavior
- Studies that offer a contribution to the social science and humanities
- Papers which are published in peer reviewed journals

The literature review included a broad range of scientific literature: action determination models; environmental behavior models; the social practices approach. This search of literature resulted in a total of more than 150 peer-reviewed studies.

Step 2: Preliminary analysis

We have grouped the papers according to different main lines: terminology; pro-environmental behavior models; and drivers and barriers. In doing so, this review aims also to complement and update previous reviews on households' energy consumption and other pro-environmental behavior models.

Step 3: Detailed analysis

A detailed analysis on both categories of energy reductions in households: the technical and behavioral energy saving measures is carried out. Followed by an overview of the most influential and commonly cited behavioral models or frameworks developed in socio-psychological research in order to provide a comprehensive explanation of energy consumption of households are described in detail, including the strengths and the weaknesses. The research topic of drivers and barriers has gained a lot of the attention of the academic community, as understanding the nature of these drivers and barriers is essential for the success of energy related policies that might encourage efficiency investments of households.

Step 4: Framework development

Based on this overview (step 3) it became clear that little research is available on what individual and social factors might influence the adoption of novel energy consumption and investment practices in households' and there is a stringent need to understand the barriers to and drivers of involvement in these. These insights and guidelines were used as a basis to build our conceptual framework on how to improve our understanding and knowledge of households' energy consumption. The framework should provide a deeper understanding in the 'what' (what factors are associated with households energy consumption, e.g., financial costs or visibility), the 'how' (how can these factors be influences, e.g., technical solutions or public policy initiatives) and the 'why' (why different types of households' are likely to behave in different context e.g., certain choices can be explained by income) in order to promote and sustain conserving practice.

3. Literature Review

3.1. Energy Efficiency in Households: Key Definitions

Various terminologies are used in the literature to describe the reduction of energy use in households. Many terms start with "energy", (energy savings, energy conservation, energy consumption, energy efficiency), while others stress more the attention on "behavior" (efficiency behavior, energy usage behavior, curtailment behavior, energy related behavior) or on the "measures" (energy saving measures, technical energy saving measures, energy efficiency measures, energy conservations measures, behavior energy saving measures) [24–27].

In order to reduce energy use in households, two broad categories of actions can be identified: "once-off actions" to save energy and "ongoing day-to-day actions" to reduce energy consumption. Once-off actions are related to efficiency behavior realized through technical energy saving measures (or energy efficiency measures). Less energy is used for a constant service, for example, an older equipment (washing machine, vehicle, etc.) replaced by a more energy efficient model (energy-efficient appliances) or investing in home improvements like insulating the roof or replacing the glazing but more efficient one. These technical measures can significantly reduce households' building and transportation energy uses and save energy and costs over long periods of time. However, they are seen as an expensive way to reduce energy consumption as they often require an initial investment. In this debate, it is also worth mentioning that, despite a growing trend to energy vulnerability of some low-income households, in Europe, energy prices (for gas, coal but also fuel for vehicles) remain relatively cheap [28,29], which led to longer return on investment for hard works such as insulation. The shift from fossil fuel to renewable energy needed to complete the international targets on CO_2 emissions should however lead to an increase in energy prices to finance this shift [30].

Day-to-day actions refer to the reduction of energy consumption through using less of an energy service as part of people's lifestyles. Turning the thermostat down a degree or two in the wintertime for example, switching off the lights, or modal shift from car to bike for short trips, etc. However, these measures are often associated with additional effort or a decrease in comfort. These behavior energy saving measures or energy conservations measures refer curtailment (energy conservation) behavior. Table 1 summarizes the main characteristics of the two previously highlighted categories of actions toward households' energy consumption.

Table 1. Energy consumption of households.

	Category 1	**Category 2**
Actions	Once-off actions	Day-to-day actions
Energy savings	Energy efficiency (Efficient energy use)	Energy conservation
Behavior	Efficiency behavior	Curtailment behavior
Strategy	Technical improvement	Different use of products and shifts in consumption
Measures	Technical energy saving measures;	Behavior energy saving measures;
	energy efficiency measures;	energy conservations measures
Amount of savings	Large energy savings	Small energy savings
Examples	Investing in home improvements e.g., insulation, energy efficient appliances, energy efficient car	Setting thermostats, switching off lights, limiting use of heating systems or car

Researchers have not been able to quantify whether efficiency behavior or curtailment behavior is more effective [24]. Some researchers have argued that curtailment behaviors initiate actual behavioral changes and sustain them for long-term [31], while others has suggested that efficiency behavior is in fact generally more effective in obtaining actual energy savings [24]. The success of the latter (efficiency behavior) may be counteracted by the "rebound-effect" (reduction in expected gains from new technologies that increase the efficiency of resource use through behavioral responses) [32].

Considering these aspects, this paper considers both categories of energy reductions in households: the technical and behavioral energy saving measures, the latter seeming somewhat overrepresented even with the knowledge that the energy saving potential of the technical measures is considered equal. The interplay between macro-level (e.g., technological innovations) and micro-level factors (e.g., use of technological innovations) will be studied in detail.

3.2. *Theoretical Framing*

Several behavioral models have been developed in socio-psychological research in order to provide a comprehensive explanation of energy consumption of households. The most influential and commonly cited frameworks are described in this section.

3.2.1. Action Determination Models

Many approaches could be categorized under the generic term of action models or action determination models. One of them is the Theory of Planned Behavior (TPB), a classical framework that has proven to be successful in explaining behavior intention and attitude in the field of household energy consumption. The TPB developed by Ajzen [33] proposes that behavior is preceded by the formation of behavioral intention. This behavioral intention depends on attitudes towards the behavior, social norms, and perceived behavioral control (the belief on whether one is capable of performing the behavior). TPB suggests that, for a specific behavior, the more active Behavioral Intention (BI) is, the more intense Subjective Norms (SN) and feel the less difficulties, individuals will be more likely to implement this behavior.

Behavioral research suggests that, values are the basis of attitude formation and it could predict behavior in a more stable and durable way than attitude [33]. In the field of environmental behavior, the Value-Belief-Norm (VBN) theory proposed by Stern et al. [34] and Stern [35] is the classical theory to study how environmental values affect the behavior. Stern divided environmental values into three dimensions: self-interest values (SV) is the belief that environmental problems will affect self-interest; altruism values (AV) is the belief that environmental issue affect others and long-term interest; biosphere values (BV) focus on natural environment intrinsic values, suggest human could not destroy the nature. The theory of VBN suggests that, environmental values are the primary antecedents to inspire public responsibility consciousness and further implement eco-environmental behavior. Another similar framework in the same line of research is the Norm-Activation Model (NAM) [36,37]. Both theories (VBN and NAM) are rooted in the thought that energy is conserved when people feel a moral obligation to do so. The VBN-theory further assumes that awareness of the problems is rooted in environmental concern and values. Thus for explaining low-cost energy curtailment behaviors, the NAM and VBN theory appeared to be successful.

However, the explanation of pro-environmental behavior is incomplete if only internal factors are considered. Guagnano, Stern et al. [38] suggest the ABC model, which incorporates the relationships of contextual factors (C), attitudes (A) and behavior (B). The ABC model involves the strategies for integrating internal processes and external conditions. Behavior is formed through the combination of personal attitudinal variables and contextual factors. Attitudinal variables include internal factors such as specific attitudes, beliefs, norms, values, information and a tendency to act in certain ways, whereas contextual factors include external factors such as physical capabilities and constraints, social institutions, legal factors and economic forces like monetary incentives and costs. The ABC model postulates that the corresponding behavior is associated with both attitudes and external conditions suggesting that behavior is an interactive product of personal-sphere attitudinal variables and contextual factors [35].

3.2.2. Social Practice Theory

Social practice theory (SPT)refers to "a routinized type of behavior which consists of several elements interconnected to one other: forms of bodily activities, forms of mental activities, 'things' and their use, a background knowledge in the form of understanding, know-how, states of emotion and motivational knowledge" [39]. It is increasingly being applied to the analysis of human behavior, particularly in the context of energy consumption. Nowadays, this theory is used as an umbrella approach under which various aspects of theory are pursued rather than a single (or specific) theory. Here the work of Shove (Lancaster University) on consumption and the group around Spaargaren (University Wageningen) on change processes is of particular relevance. The primary insights focusses not on individual behavior but on social practice and on the interaction of people's practices and in particular their material contexts. This leads towards reflecting upon why certain practices are done, and how and why other practices are prevented. Shove stresses the importance on how social practice have changed over time, how it becomes normal and what the consequences on sustainability are. She is doing this using the concepts of cleanliness, comfort and convenience [40,41]. Spaargaren uses Shove's theoretical approach and place the social practices into a conceptual model, which has a strong emphasis on sustainability of existing lifestyles and on the ecological modernization of the society [42].

3.2.3. Integrated Perspectives

Nowadays, the conducted studies seem to focus more on the interaction of multiple factors, the integrating of different theories/perspectives and the multiplicity of forces underpinning energy consumption and conservations. Venkatesh [43] proposed the Unified Theory of Acceptance and Use of Technology (UTAUT), a synthesis of eight existing models of technology acceptance. The model integrates elements from Theory of Reasoned Action (TRA), Motivational Model (MM), Theory of Planned Behavior (TPB), Technology Acceptance Model (TAM), a combined Theory of Planned Behavior/Technology Acceptance Model (C-TPB-TAM), Model of PC Utilization (MPCU), Innovation Diffusion Theory (IDT), and Social Cognition Theory (SCT).

Turaga et al. [44] integrated for example the moral considerations of VBN with the rational framework of TPB and Bamberg [45] combined the TPB and the NAM. Abrahamse et al. [24] proposed that both micro-level factors and macro-level factors can influence household energy consumption. And, some researchers have investigated different types of energy consumer profiles in order to pinpoint what specific factors are associated with energy-saving behavior, e.g., Guerra Santin [46] and Gaspar and Antunes [47].

Table 2 summarizes the main behavior change models used in energy research. The name of the model and its principal proponent is given in Table 2, followed by some strengths and weaknesses. Due to the restricted length of this paper, it will not have been possible to describe every single facet of each model.

Table 2. Strengths and weaknesses of behavior change models.

Constructs	Main Concept and Strengths	Weaknesses	Empirical Evidence
Attitude-Behavior-Context Model (ABC model)—Stern & Oskamp [48]			
Attitude; Behavior; Context	Behavior (B) is an interactive outcome of personal attitudinal variables (A) and contextual (C) factors.	Does not take into account the influence of habits.	[49]
Consumption as Social practice theory—Spaargaren & Van Vliet [42]			
Social practices; Lifestyle; System of provision; Consumption	Describes a mutual dependency between domestic consumers and external systems that provides domestic goods where consumers are unable to engage in environmentally sustainable lifestyles unless external systems provide facilitative goods and take into account consumers' domestic practices	Difficulty of defining exactly what a practice is.	[50]
Diffusion of Innovation (DI)—Rogers [51]			
Innovation; Communication channels	Explain the process by which people adopt a new idea, behavior or object. It specifies numerous mechanisms through which adoption is achieved, and factors that influence the choices an individual makes.	The theory does not consider the possibility that people will reject an innovation even if they fully understand it. Does not take into account that adoption of new technologies is constrained by situational factors (lack of resources or access to these technologies).	[52–54]
Goal-Framing Theory—Lindenberg & Steg [55]			
Hedonic goals; Gain goals; Normative goals	Propose that goals direct the iformation and cognitions that people attend to. It proposes three types of goals (hedonic, gain and normative), an states that activation of each type directs people's attention to different sub goals, cognitions and information	The three goals are not equal of strength	
Model of pro-Environmental behavior—Kollmuss & Agyeman [56]			

Table 2. *Cont.*

Constructs	Main Concept and Strengths	Weaknesses	Empirical Evidence
Internal and external factors; Barriers;	This theory is composed by internal and external factors that can contribute to environmentally friendly behavior, alongside a number of barriers to pro-environmental behavior.		
Norm Activation Theory (NAT)-Schwartz [36]			
Activation of norms; Perception of need and responsibility; Assessment, evaluation and reassessment; action or inaction response	Explain the decision making process underlying altruistic and environmentally friendly behavior.	Intention, past experience and habit as factors influencing altruistic behavior are not considered.	[57–59]
Self-Determination Theory—Deci & Ryan [60]			
Intrinsic and extrinsic motivation	Comprising 5 theories and provide a broad framework to study motivation, personality and behavior.	Only when individuals are intrinsically motivated towards an activity is the behavior considered to be fully self-determined.	[61]
Social Learning Theory—Miller & Dollard [62]			
Drive; Cue; Response; Reward	Explains how imitative learning takes place, with four factors instrumental to the learning process.		
Stage model of self-regulated behavior change—Bamberg [45]			
Self-regulating process; Goal intention; Behavioral intention; Implementation intention	The perception of individual responsibility and negative effects of personal behaviors activate social norms and thus could lead to behavioral change.	Not include external factors and past experience.	[63,64]
	It describes the behavioral change process and the individual intention or willingness to change behavior toward a pro-environmental behavior by four stages.	Based on the assumption that it is possible for people to move up the ladder of sustainable behavior.	
Theory of Interpersonal behavior—Triandis [65]			
Behavior intention; Habits; Social factors	Intentions, and habits, influence behavior, which are also affected by facilitating conditions (external factors).	Has not been as widely used in empirical	[57]
Theory of Reasoned Action (TRA)—Ajzen & Fishbein [66]			
Attitude; Subjective norms; Intention; Behavior	Relationship between attitudes and behaviors within human action.	Issues such as cognitive deliberation, habits and the influence of affective or moral factors are not addressed.	[67]
		Unable to account for behaviors not under volitional control.	
Theory of Planned Behavior (TPN)—Ajzen [33]			

Table 2. *Cont.*

Constructs	Main Concept and Strengths	Weaknesses	Empirical Evidence
Attitude; Subjective norms; Perceived behavioral control; Intention; Behavior	Builds on the TRA model and includes a new determinant of perceived behavioral control to predict behavior - person's belief on how difficult or easy a behavior will be influences his/her decision to conduct that behavior.	Experience is not included in the model, so the measurement of actual behavior is missing. Personality characteristics, demographic variables and factors such as social status are excluded from the model.	[57,68–70]
The Social Practice Theory [40,41]			
Comfort; Cleanliness; Convenience	Introduces three domains of everyday life, those of comfort, cleanliness, and convenience (CCC). By using these concepts, Shove explores the questions of how new conventions become normal, and what the consequences are for sustainability	A group of individuals is seen as one single actor instead of all the individuals represented as such	[71]
Value Belief Norm Theory (VBN)—Stern et al. [34]			
Personal values; New ecological paradigm; Awareness of consequences; Ascription of responsibility; Personal moral norms;	Explains environmentalism and conservation behavior. It proposes a casual chain of values, beliefs and norms that leads to support for a social movement	Each variable in the chain affects the next and may variables more than one level down the chain. Thus all variables have to be analysed to identify the most influential factors. Values often fail to predict specific behaviors and weak correlation between personal norm and indicators of pro-environmental behavior	[72]
UTAUT 1,2,3—Davis; Venkatesh and& Davis [73,74]			
Perceived usefulness; Perceived ease of use; Intention; Subjective norm	It reviews eight models. Describes the factors that influence the acceptance/usage of technology, and the mechanisms underlying these influences. Central to the model is the proposal that acceptance is determined by two factors, namely perceptions of ease of use and perceptions of usefulness.	It is considered a less parsimonious theory	[74,75]

For some behavior change models we do not have indicated any weaknesses or relevant empirical evidence (see empty cells in the table).

The diversity and variety of the behavioral models or theories have shown that pinpointing the right type of constructs/indicators to achieve behavioral change is not straightforward. Jackson ([49]) sums up this problem in his discussion of consumer behavior. *"Beyond a certain degree of complexity, it becomes virtually impossible to establish meaningful correlations between variables or to identify causal influences on choice. Conversely,... simpler models run the risk of missing out key causal influences on a decision, by virtue of their simplicity... this means that there will always be something of tension between simplicity and complexity in modelling consumer behavior. More complex models may aid conceptual understanding but be poorly structured for empirical quantification of attitudes or intentions (for example). Less complex models may aid in empirical quantification but hinder conceptual understanding by omitting key variables or relationships between key variables".*

Behavior is a complex combination of different constructs/indicators (values, norms, habits, social factors) and changing any of these can be challenging. Last but not least, it is worth mentioning that there is not a framework that is universally accepted by scholars as providing a comprehensive explanation of households' energy consumption and conservation.

Although the overview provided in this paper does not intend to be exhaustive and the selected models vary in purpose and context, the following insights and guidelines can be highlighted and used as a basis to build our conceptual framework:

- Consistent terminology for key constructs, some models used different terms interchangeably for the same construct
- Focus on current behavior rather than generating behavior change, given that most models use static data
- The importance of considering motivation, ability and barriers arising from the physical and social environment as important factor
- Concept of social norms was brought in the models in slightly different ways, in some models differentiations were made between different types of social norms
- Behavior change involves going through a series of stages (stage-based approaches), however, we have not found an advantage over other (more dynamic) models

3.3. Households' Energy Consumption: Key Determinants

Much research has been conducted over the years to clarify the key determinants that influence households' energy consumption. They may have been differing motives as to why research has looked at this domain, however, the overarching aim has been the focus on the reduction of energy consumption. Whether it is considered from an economic perspective (household's energy consumption linked to and have monetary impacts) or from a perspective related to environmental impacts does not matter. This section provides an overview of recent developments in the literature with regard to factors influencing households' energy consumption. Non-residential buildings are out of the scope of this paper and therefore literature in this field is not considered.

A classification of the identified influencing factors underlying this behavior in the residential sector is proposed to identify the determinants affecting households' energy consumption. Gärling et al. [76] argued that in order to change people's environmental behavior there is a need to consider both macro and micro-level factors. Jackson [49] divided all the influencing factors into internal factors (including attitudes, beliefs and norms) and external factors (including regulations and institutions). This paper follows his classification line and examines the following classification for the factors as possibly affecting energy-saving behavior: internal level factors, external level factors and social factors. Regarding the latter, previous research [41,49] has come up with useful conclusions that the social embeddedness/ social context is understudied. More in detail, how individual choices are continually being shaped and reshaped by the social contexts is important to consider in this research.

3.3.1. Internal Level Factors Influencing Household Energy Consumption

Various internal level factors influence household energy use and energy savings. Steg and Vlek [77], one of the most relevant publications on residential energy behaviors, identified motivational factors, contextual factors and habitual behavior as the most important factors in environmental behavior.

Motivational factors are defined as subjective individual characteristics that may influence how people perceive and rate the acceptability of objective characteristics of energy alternatives. *Habitual factors* refer to individual factors habitual and guided by automated cognitive processes rather than being preceded by elaborated reasoning. How these habits are formed, reinforced and sustained is important for designing effective interventions to modify this behavior. *Contextual factors* refers to the objective characteristics of energy alternatives determined by its own context for example the energy price.

3.3.2. External Level Factors Influencing Household Energy Consumption

The second group of identified influencing factors place behavior as a function of processes and characteristics external to the individual, these include amongst other fiscal and regulatory incentives, and institutional constraints. In the literature, the term 'institution´ can play different roles in transition trajectories/ innovations and various authors [78–80] do not mean the same things when using this term. More in detail, Lundvall consider institutions as 'things that pattern behavior' (such as norms, rules, and laws), while Nelson and Rosenberg institutions consider as 'formal structures with an explicit purpose' (often called as organizations) [79]. In this research, the term institutional factors are used to describe the rules, regulations, standards and so on that shapes the behavior of households in terms of perceptions and actions. Institutional change can therefore greatly influence how households perceive and respond to uncertainties in the energy usage.

3.3.3. Social Factors Influencing Household Energy Consumption

The effect of social interaction on energy-saving behavior is also emphasized in some studies [81–86]. Social norm and social identity studies in the energy domain have generally looked at their influence on consumption patterns and have showed their effectiveness when used in intervention studies to reduce energy consumption. As social norms signal what the members of the communities we live in do, as well as what they approve or disapprove off, they are an important determinant of individual behavior both at home and on the road.

Furthermore, the importance of considering the group membership as an indicator of the importance of cultural contexts and social influences on consumer behavior has also been identified in previous research [87–90]. Individuals with a strong sense of group membership (i.e., with a high group identification), typically express positive evaluations, display the tendency to act in favor, and strive to maintain a positive image of their in group, even at the expense of an out-group. Social psychological studies showed social identity as one of the main psychological factors leading to voluntary cooperation to solve commons problems or dilemmas by postponing their narrow self-interest and to act on behalf of their group, community or place.

Table 3 gives an overview of the most commonly identified influencing factors correlated with household energy consumption, both energy efficiency and energy conservation.

Table 3. Overview of the key literature that identify the factors affecting households' energy consumption.

Category	Characteristics	Literature
Internal context	Socio-demographic, Contextual factors, Attitudes, behaviors and habits (implicit behavior)	[19,24,26,31,58,59,72,82,88,91–120]
External context	Incentives, institutional, and infrastructures	[24,67,101,121–130]
Social context (internal and external factors)	The role of social norms, social identity and social practices (incl. social systems)	[59,69,81,82,94,131–136]

The concepts of "day-to-day actions" and "once-off actions" presented in Chapter 2 are particularly used in this vision. Household's energy saving behavior indeed includes a wide range activities from habitual day-to-day actions to sophisticated and costly once-off actions [27]. That is why it should be noted that the above determinants of household's energy conservation behavior affect these various types of activities in different ways depending on type of behavior and involvement with the product and behavior and have different psychological properties [99].

While once-off actions are one-time purchase decisions characterized though initial financial expenses and the potential for future savings, curtailment behaviors or ongoing day-to-day actions are considered to be routinized or habitual in the sense that it spares individual's time and effort of decision-making on issues that re-occur regularly [90,119]. In comparison to one-time purchase decisions that might have the side effect of increasing consumer's comfort, day-to-day actions implicate additional effort or decreased comfort.

Psychological factors including values, beliefs, attitudes and norms have been identified to be successful in predicting curtailment behavior [117,137]. For example, personal norms affect both curtailment behavior and involvement in purchase decisions through feeling a moral obligation to do so. This is also the case for environmental beliefs in the form of ascription of responsibility [118]. Eriksson et al. [138] and Nordlund and Garvill [139] have shown in their research on car use that there is a strong influence of personal norms for the willingness to curtail personal car use.

In general, Gatersleben et al. [137] and Whitmarsh [117] delivers an empirical evidence that daily energy saving actions are more likely to be influenced by internal factors, while actions which require considerable monetary costs (energy efficiency investments) are more dependent on guided circumstances. However, Jansson and Marell [118] shows in their empirical research that for both high involvement once-off actions and ongoing day-to-day actions biospheric values and personal norms have a strong influence on their energy reduction.

Regarding socio-demographic factors such as age, living status and gender, existing literature provides evidence both for and against hypothesis in either direction. Lee et al. [140] show that there are some gender differences in adoption of energy-efficient lighting at home in the sense that women are more likely to adopt energy-saving practices and were more willing to pay a higher price for energy-efficient light sources. Poortinga et al. [26] show that couples and families found technical efficiency measures more acceptable than singles did. According to Sardianou [112], energy saving investments are less likely to be made by older households since these households believe in shorter stream of benefits from energy improvements than other age cohorts. Another explanation is that younger households prefer an up-to-date technology which is most of the time also more efficient, while older households accept their older appliances and replace them only when necessary [114]. Carlsson-Kanyama et al. [114] also prove that households with younger head of the family are more likely to adopt energy-saving measures. However, Guerin et al. [120] show that age and the energy saving curtailment behavior is positively correlated. Poortinga et al. [26] also provides empirical evidence for the hypothesis that energy efficiency measures are more acceptable for households with a high income, while behavioral energy saving measures aimed at reducing direct energy costs were the least acceptable for high incomes. This might be explained, as seems to be straightforward, by the fact that energy efficiency measures (technical measures) often require an initial investment, which seems to be less problematic for households with a high income [112]. Another possible explanation for this phenomenon is the fact that day-to-day actions implicate a decrease in comfort while one-of actions might even increase consumer's comfort. Stern and Gardner [141], show that the home ownership also causes differences between households, energy efficiency investments is meaningful for homeowners whereas curtailments might be the only option for renters.

3.3.4. Discussion

A number of key determinants have been identified in the literature, ranging from situational factors in the external environment (e.g., contextual, structural and institutional factors) through to more person-specific attributes of consumers themselves (e.g., socio-demographic, psychological factors). Despite an expanding literature, empirical evidence of the impact of the latter two broad categories of variables that have been identified, socio-demographic factors (e.g., income, employment status, dwelling type/size, home ownership, household size) and psychological factors (e.g., beliefs and attitudes, motives and intentions, perceived behavioral control, cost-benefit appraisals, personal and social norms) has not been consistent and conclusive to date. However, a common finding that has been well documented by behavioral economists, psychologists and other social scientists is that individuals do not always behave more sustainably despite having positive attitudes or behave logically to favorable economic choices in order to reduce household energy consumption.

Another common identified finding, combining financial incentives with program components (like energy assessments, information, education, appeals, informal social influences, convenience and quality assurance) reduce the transaction costs of targeted/ desired actions and have shown synergistic effects greater than the additive effects of individual interventions or policy. Furthermore, previous research has shown the importance of the full range of consistent knowledge of the environmental, economic and social impact for policy makers and financing institutions to decide whether or not to support new business models. For example, smart metering has been widely pushed, despite little knowledge on the environmental impacts as well as social impacts such as data security.

But both strands of action (one-off investment action or continuous action) require important and coordinated changes in household practices that go beyond passive assumption of energy-efficient technologies and acceptability of traditional policy measures. Efforts to change household energy use through information campaigns have proven very limited [24,129,142] and recent trends in diversification of energy generation and changing consumer roles have underlined the potential for smarter transformation potentials in harnessing the active households [143–145]. Nevertheless, little research is available on what individual and social factors might influence the adoption of novel energy consumption and investment practices in households and there is a stringent need to understand the barriers to and drivers of involvement in these. The challenge is to understand the internal, social and external level factors that threaten the energy use in household, so that energy-saving behaviors could be facilitated. Furthermore, the effects of contextual factors on energy usage behavior need to be studied in more detail, as well as how these factors might be affect various environmental and motivational factors. This in turn should lead to an extension of the existing methodological and/or theoretical models.

4. A Framework to Understand Household's Energy Consumption

Based on the literature review provided in the previous section, the identified key influencing factors are summarized as possibly affecting energy-saving behavior in a conceptual framework, presented on Figure 1. Whether household energy consumption is based on a one-off investment action or continuous actions, behavior is influenced by the external as well as by the internal context. External context such as institutional factors, technological developments, economic growth, cultural developments influence behavior at the broader level, while attitudinal and personal factors such as demographic factors and motivations shape behavior at the individual level. To illustrate the framework, the dimension of the external context, the internal context-attitudinal factors and the internal context-personal factors is connected. In order to differentiate between determinants of a different nature, a distinction between contextual, economic and social variables is proposed. These have different positions in the model and operate at different levels of influence. Regarding these levels of influence, personal (household) factors at the level of internal context have the biggest influence on the energy-saving behavior of households in contrast to the factors at the level of external context. A personal level economic factor can be altered relatively quickly, for example by a change in income, while the introduction of a subsidy program for all households in a country is often over the course of a few years.

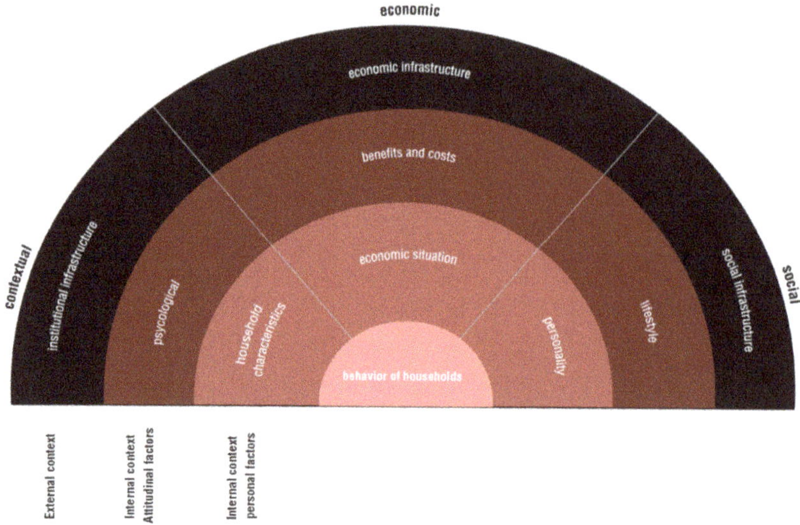

Figure 1. Conceptual framework to understand households' energy efficiency behavior.

Internal Context—Personal Factors

As already explained, the characteristics of each household have a direct impact and the biggest influence on the behavior of households. Their characteristics reflect the attitude and experiences from a descriptive angle (e.g., ownership, size and type). The availability of the requirements necessary to adopt technical and behavioral measures (economic situation) is also an important issue. This obviously affects what a household can afford, but the perspective on money and the level of importance of price in certain purchasing decisions does not belong to personal factors but has to do with the attitude people have. The personality of the people is clustered in the social context, however, it is composed by role and status, age or gender which represents a strong connection for certain behavior.

Internal Context—Attitudinal Factors

Attitudinal factors include factors held by the individual that affect the choices and the behavior people undertakes. These include an individual's motivation (e.g., pro-social, altruistic), perception, beliefs and attitudes which are part of the contextual process affecting the individual intentions. It includes also calculations which people make before acting, including personal evaluations of costs and benefits. Thus in spite of the advantage of adopting more efficient appliances, the cost of that decision has to be in concordance with the perceived benefits. Even when energy-saving measures are affordable, the balance between costs and benefits could represent a major barrier due to the uncertainty. For instance, the fact of thinking about long-term benefits when costs are immediately perceived has a direct effect in the attitudinal factor regarding to the intention behavior (especially in the case of the adoption of photovoltaic panels, as shown in [99,146,147].

The lifestyle of people such as group membership, normative social influence and family are also important factors. The indirect commitment with society makes behavior to be on the same line with the others tended to follow social system flow. Also one has to cooperate with other household members.

External Context

The external environment comprises situational opportunities and dependence of other. It can be interpret as set of regulations, system of laws, political environment and governance structure which interrelated control the distribution and consumption of energy adopting new measures in households.

At the social context for example, one has to obtain correct information about the most effective ways of reduction in order to reduce household energy consumption.

Table 4 provides an overview of the three categories of potentially important variables that have been identified for explaining variability in energy reduction of households: contextual, economic and social variables. These variables are divided over different levels with an explanation of the determinants of household energy behavior.

Table 4. Determinant of household energy consumption.

Level of Factors	General Determinants	Detailed Determinants
External environment		
Contextual	Institutional infrastructure	Laws, regulations and policies Availability technology Built environment (Infrastructure)
Economic	Economic infrastructure	Pricing (tariffs, rebates and subsidies) Information, mass media and advertising
Social	Social infrastructure	Neighborhood factors (community spirit and community norms) Broader public norms
Internal environment—Attitudinal factors		
Contextual	Psychological factors	Motivation Perception Beliefs and attitudes Knowledge and awareness (learning)
Economic	Benefits and costs	Energy consumption pattern Financial cost Benefit appraisal (potential impact of cost)
Social	Lifestyle	Group membership Normative social influence Family
Internal environment—Background factors		
Contextual	Household characteristics	Size and type Dwelling (ownership, age, size) Geographical locations (region, rural-urban, climate)
Economic	Economic situation	Income Employment status Education
Social	Personality	Role and status Age Gender

In summary, the conceptual model shows that energy consumption of households is based on a complex interaction between contextual, economic and social influence. This interaction has been structured into three categories implying a multilevel division of factors to shape the process of households' behavior and its transition to assume and adopt new insights affecting their day-to-day actions. The conceptual framework suggests a range of determinants for energy-saving behavior at different levels. However, it should be noted that an important point of attention is which specific label to be used in the conceptual framework and where the specific labels should be placed. This could be related to the disciplinary angle from which one approaches the framework. This is especially the case along the boundary of the social context. Although all the determinants are presented separately, from a practical approach are working synergistically and interrelated influencing the behavior and their current performance in households.

The framework is not only interesting for researchers, but also for policymakers (at the national and local level), practitioners (energy providers and engineers), as well as for social energy networks. First, it is interesting for policymakers in the area of energy provision for households, at national and local levels. At a national level, the gained insights into the "what", the "how" and the "why" provides handholds to formulate an appropriate policy or service view that can help the government to transform the current energy system into a more sustainable one. In order to motivate these households, 'education and communication' is an important issue. Education on the interrelated issues of energy, climate change, and sustainability, and communication of strategies for reducing consumption and emissions (ranging from energy efficiency and conservation to more sustainable energy technologies). At a local level, households (and communities) can start participating more in bottom-up energy initiatives, thereby increasing the share of more sustainable energy technologies in the energy market. These results are also helpful for local governments and their planners as they have an important role to play in promoting more sustainable energy technologies. But, for both, national and local policymaker, these insights stresses the importance of creating policies that are transparent and easy to take advantage of. Second, we find that trustworthy information about the contextual (e.g., performance) and economic (e.g., costs) dimension is an important factor influencing interest in speaking with practitioners like energy providers and engineers. And finally, that households may seek such information from the experiences of personal connections in their neighborhoods and social networks (social dimension).

5. Conclusions and Perspectives for Further Research

Our intention in the paper has been to introduce a new conceptual framework to capture and understand households' energy consumption, efficiency behavior and curtailment behavior. Households have been progressively highlighted as the main actor that can play a substantial potential in the reduction of this energy use. Their behavior is a complex and hard to change process that combines numerous determinants. These determinants have already been extensively studied in the literature from a wide range of thematic areas each by its own assumptions and often neglect important energy behavioral components, therefore, energy behavior studies require an integration of disciplines through an interdisciplinary approach. Based on that knowledge, this paper aims at connecting objective (physical and technical) with subjective (human) aspects related to energy use of households in one framework. This combination should provide the answers to the 'what', the 'how' and most importantly the 'why' questions about people's behavior regarding energy use. This proposed framework allows clarifying the numerous internal and external factors that act as key determinants, as well as the need to take into account their interactions. Moreover, it would re-form demand as one of the result of interactions in and between the contextual, economic and social contexts in which households' lives. It would, however, not obviate the individual household nor research that intended to track changes in how individual households think and act. The framework proposed in this paper opens avenues for the integrated study of households' energy consumption and has further potential policy implications to better capture and take into account behaviors in policies, incentives and regulations still often focused on technical aspects.

Further studies are suggested to use the proposed framework for explaining households energy behavior focusing on identifying the specific factors that influence household energy usage (e.g., consumption) and changes in energy use over time (e.g., curtailment and efficiency behaviors). The framework has to be applied to an increasing set of empirical cases (for example PV and LED) carried out in a way as to systematically explore the opportunities and barriers, which in turn can enhance our understanding of how determinants interact as part of a larger explanatory framework.

Author Contributions: Conceptualization, V.V., A.-F.M. and V.U.; methodology, V.V. and A.-F.M.; validation, V.V., A.-F.M. and V.U.; formal analysis, V.V.; investigation, V.V.; resources, V.V.; data curation, V.V.; writing—original draft preparation, V.V.; writing—review and editing, A.-F.M. and V.U.; visualization, V.V.; funding acquisition, V.V.

Funding: This research is part of the research programme JSTP—Joint Research Projects: Smart Energy in Smart Cities, titled 'Energy efficiency of households in cities. A multimethod analysis' funded by the Netherlands Organisation for Scientific Research (NWO), grant number 467-14-023. Furthermore, the research presented in this paper also received funding from the European Union's H2020 Research and Innovation program under grant agreement number 727642. The sole responsibility for the content of this paper lies with the authors.

Conflicts of Interest: The authors declare no conflict of interest.

References

1. Brounen, D.; Kok, N.; Quigley, J.M. Energy literacy, awareness, and conservation behaviour of residential households. *Energy Econ.* **2013**, *38*, 42–50.
2. Energy Information Administration (EIA). International Energy Outlook 2016. *Indep. Stat. Anal.* **2016**. Available online: https://www.eia.gov/outlooks/ieo/pdf/0484 (accessed on 6 November 2019).
3. Nejat, P.; Jomehzadeh, F.; Taheri, M.M.; Gohari, M.; Majid, M.Z. A global review of energy consumption, CO2 emissionsand policy in the residential sector (with an overview of the top ten CO2 emitting countries). *Renew. Sustain. Energy Rev.* **2015**, *43*, 843–862.
4. Jones, P.J.; Lannon, S.; Williams, J. Modeling building energy use at urban scale. In Proceedings of the 7th International IBSPA Conference, Rio de Janeiro, Brazil, 13–15 August 2001; pp. 175–180.
5. Marique, A.F.; Reiter, S. A simplified framework to assess the feasibility of zero-energy at the neighbourhood / community scale. *Energy Build.* **2014**, *82*, 114–122.
6. Ratti, C.; Baker, N.; Steemers, K. Energy consumption and urban texture. *Energy Build.* **2005**, *37*, 762–776.
7. Steemers, K. Energy and the city: Density, buildings and transport. *Energy Build.* **2003**, *35*, 3–14.
8. Johansson, T.B.; Patwardhan, A.P. *Global Energy Assessment: Toward a Sustainable Future*; Cambridge University Press: Cambridge, UK, 2012.
9. Marique, A.F.; Reiter, S. A Method to Evaluate the Energy Consumption of Suburban Neighbourhoods. *HVACR Res.* **2012**, *18*, 88–99.
10. European Commission. *A Roadmap for Moving to a Competitive Low Carbon Economy in 2050*; Communication from the Commission to the European Parliament, the Council, the European Economic and Social Committee and the Committee of the Regions; European Commission: Brussels, Belgium, 2011.
11. Lysen, E. The Trias Energica: Solar Energy Strategies for Developing Countries. In Proceedings of the EUROSUN Conference, Freiburg, Germany, 16–19 September 1996.
12. Hamilton, I.G.; Steadman, J.P.; Bruhns, H.; Summerfield, A.J.; Lowe, R.J. Energy efficiency in the British housing stock: Energy demand and the Homes Energy Efficiency Database. *Energy Policy* **2013**, *60*, 462–480.
13. Hong, S.H.; Gilbertson, J.; Oreszczyn, T.; Green, G.; Ridley, I. Warm Front Study Group. A field study of thermal confort in low-income dwellings in England before and after energy efficient refurbishment. *Build. Environ.* **2009**, *44*, 1228–1236.
14. Ma, Z.; Cooper, P.; Daly, D.; Ledo, L. Existing building retrofits: Methodology and state-of-the-art. *Energy Build.* **2012**, *55*, 889–902.
15. McLaren, D. Energy poverty and the future of urban retrofit. In *Urban Retroffiting for Sustainability: Mapping the Transition to 2050*; Dixon, T., Eames, M., Lannon, S., Hunt, M., Eds.; Routeledge: Oxon, UK, 2014.
16. Nemry, F.; Uihlein, A.; Colodel, C.M.; Wetzel, C.; Braune, A.; Wittstock, B.; Hasan, I.; Kreißig, J.; Gallon, N.; Niemeier, S.; et al. Options to reduce the environmental impacts of residential buildings in the European Union—Potential and costs. *Energy Build.* **2010**, *42*, 976–984.
17. Office of Climate Change. *Household Emissions Project*; Final Report; Office of Climate Change: Canberra, Australia, 2007.
18. De Meester, T.; Marique, A.F.; De Herde, A.; Reiter, S. Impacts of occupant behaviours on residential heating consumption for detached houses in a temperate climate of the northern part of Europe. *Energy Build.* **2013**, *57*, 313–323. [CrossRef]
19. Lopes, M.A.R.; Antunes, C.H.; Martins, N. Energy behaviours as promoters of energy efficiency: A 21st century review. *Renew. Sustain. Energy Rev.* **2012**, *16*, 4095–4104. [CrossRef]
20. Santin, O.G.; Itard, L.; Visscher, H. The effect of occupancy and building characteristics on energy consumption for space and water heating in Dutch residential stock. *Energy Build.* **2009**, *4*, 1123–1232.
21. Jones, R.V.; Fuertes, A.; Boomsma, C.; Pahl, S. Space heating preferences in UK social housing: A socio-technical household survey combined with building audits. *Energy Build.* **2016**, *127*, 382–398. [CrossRef]
22. Wei, J.; Jones, R.; de Wilde, P. Driving factors for occupant-controlled space heating in residential buildings. *Energy Build.* **2014**, *70*, 36–44. [CrossRef]
23. Yin, R.K. *Case Study Research—Design and Methods*; SAGE Publications: Thousand Oaks, CA, USA, 2003.
24. Abrahams, W.; Steg, L.; Vlek, C.; Rothengatter, T. A review of intervention studies aimed at household energy conservation. *J. Environ. Psychol.* **2005**, *25*, 273–291. [CrossRef]

25. Frederiks, E.R.; Stenner, K.; Hobman, E.V. The Socio-Demographic and Psychological Predictors of Residential Energy Consumption: A Comprehensive Review. *Energies* **2015**, *8*, 573–609. [CrossRef]
26. Poortinga, W.; Steg, L.; Vlek, C.; Wiersma, G. Household preferences for energy-saving measures. A Conjoint Analysis. *J. Econ. Psychol.* **2003**, *24*, 49–64. [CrossRef]
27. Urban, J.; Scasny, M. Exploring domestic energy-saving: The role of environmental concern and background variables. *Energy Policy* **2012**, *47*, 69–80. [CrossRef]
28. Audenaert, A.; Timmerman, V. A cost-benefit model to evaluate energy saving measures in office buildings. *WEAS Trans. Environ. Dev.* **2011**, *12*, 371–384.
29. Carlini, M.; Zilli, D.; Allegrini, E. Simulating building thermal behaviour: The case study of the school of the stae forestry corp. *Energy Procedia* **2015**, *81*, 55–63. [CrossRef]
30. Fabra, N. Towards a Low Carbon European Ower Sector. In *The Energy Transition in Europe: Initial Lessons from Germany, the UK and France*; Centre on Regulation in Europe (Cerre): Brussels, Belgium, 2015.
31. Geller, E.S. The Challenge of Increasing Proenvironment Behavior. In *Handbook of Environmental Psychology*; Bechtel, R.G., Churchman, A., Eds.; Wiley: New York, NY, USA, 2002; pp. 525–540.
32. Ehrhardt-Martinez, K.; Laitner, J.A. Rebound, Technology and People: Mitigating the Rebound Effect with Energy-Resource Management and People-Centered Initiatives. In Proceedings of the ACEEE Summer Study on Energy Efficiency in Buildings, Washington, DC, USA, 15–20 August 2010; Available online: http://www.aceee.org/files/proceedings/2010/data/papers/2142.pdf (accessed on 6 November 2019).
33. Ajzen, I. The theory of planned behavior. *Organ. Behav. Hum. Decis. Process.* **1991**, *50*, 179–211. [CrossRef]
34. Stern, P.C.; Dietz, T.; Abel, T.; Guagnano, G.A.; Kalof, L. A value-belief-norm theory of support for social movements: The case of environmental concern. *Hum. Ecol. Rev.* **1999**, *6*, 81–97.
35. Stern, P.C. New Environmental Theories: Toward a Coherent Theory of Environmentally Significant Behavior. *J. Soc. Issues* **2000**, *56*, 407–424. [CrossRef]
36. Schwartz, S. The Justice of Need and the Activation of Humanitarian Norms. *J. Soc. Issues* **1975**, *31*, 111–136. [CrossRef]
37. Schwartz, S.H.; Howard, J.A. A normative decision-making model of altruism. In *Altruism and Helping Behavior*; Rushton, J.P., Sorrentino, R.M., Eds.; Erlbaum: Hillsdale, MI, USA, 1981; pp. 89–211.
38. Guagnano, G.A.; Stern, P.C.; Dietz, T. Influences on Attitude-Behaviour Relationships: A Natural Experiment with Curbside Resycling. *Environ. Behav.* **1995**, *27*, 699–718. [CrossRef]
39. Reckwitz, A. Toward a Theory of Social Practices. A Development in Culturalist Theorizing. *Eur. J. Soc. Theory* **2002**, *5*, 243–263. [CrossRef]
40. Shove, E. Converging Conventions of Comfort, Cleanliness and Convenience. *J. Consum. Policy* **2003**, *26*, 395–418. [CrossRef]
41. Shove, E. *Comfort, Cleanliness and Convenience: The Social Organization of Normality*; Berg: Oxford, UK, 2003.
42. Spaargaren, G.; van Vliet, B. Lifestyles, Consumption and the Environment; the Ecological Modernization of Domestic Consumption. *Environ. Politics* **2000**, *9*, 50–76. [CrossRef]
43. Venkatesh, V.; Morris, M.; Davis, G.B.; Davis, F.D. User Acceptance of Information Technology: Toward a Unified View. *MIS Q.* **2003**, *27*, 425–478. [CrossRef]
44. Turaga, R.M.; Howarth, R.B.; Borsuk, M.E. Pro-environmental behavior: Rational choice meets moral motivation. *Ann. N.Y. Acad. Sci.* **2010**, *1185*, 211–224. [CrossRef] [PubMed]
45. Bamberg, S. Changing environmentally harmful behaviors: A stage model of self-regulated behavioral change. *J. Environ. Psychol.* **2013**, *34*, 151–159. [CrossRef]
46. Guerra Santin, O. Behavioural patterns and user profiles related to energy consumption for heating. *Energy Build.* **2011**, *43*, 2662–2672. [CrossRef]
47. Gaspar, R.; Antunes, D. Energy efficiency and appliance purchases in Europe: Consumer profiles and choice determinants. *Energy Policy* **2011**, *39*, 7335–7346. [CrossRef]
48. Stern, P.; Oskamp, S. Managing scarce environmental resources. In *Handbook of Environmental Psychology*; Altman, I., Stokels, D., Eds.; Wiley: New York, NY, USA, 1987; pp. 1044–1088.
49. Jackson, T. Motivating Sustainable Consumption, a review of evidence on consumer behaviour and behavioural change. *Sustain. Dev. Res. Netw.* **2005**, *29*, 30. Available online: http://sustainablelifestyles.ac.uk/sites/default/files/motivating_sc_final.pdf (accessed on 6 November 2019).

50. Spaargaren, G.; Beckers, T.; Martens, S.; Bargeman, B.; van Es, T. *Gedragspraktijken in Transitie. De Gedragspraktijkenbenadering Getoetst in Twee Gevallen: Duurzaam Wonen en Duurzame Toeristische Mobiliteit*; Globus—Wageningen University Environmental Sciences: Tilburg, The Netherlands, 2002.
51. Rogers, E. *Diffusion of Innovations*; Free Press: New York, NY, USA, 1995.
52. Faiers, A.; Neame, C. Consumer attitudes towards domestic solar power systems. *Energy Policy* **2006**, *34*, 1797–1806. [CrossRef]
53. Kaplan, A.W. From passive to active about solar electricity: Innovation decision process and photovoltaic interest generation. *Technovation* **1999**, *19*, 467–481. [CrossRef]
54. Labay, D.G.; Kinnear, T.C. Exploring the consumer decision process in the adoption of solar energy systems. *J. Consum. Res.* **1981**, *8*, 271–278. [CrossRef]
55. Lindenberg, S.; Steg, L. Normative, gain and hedonic goal-frames guiding environmental behavior. *J. Soc. Issues* **2007**, *63*, 117–137. [CrossRef]
56. Kollmuss, A.; Agyeman, J. Mind the Gap: Why do people act environmentally and what are the barriers to pro-environmental behaviour? *Environ. Educ. Res.* **2002**, *8*, 239–260. [CrossRef]
57. Bamberg, S.; Schmidt, P. Incentives, Morality or Habit: Predicting students' car use for university routes with the models of Ajzen, Schwartz and Triandis. *Environ. Behav.* **2003**, *35*, 264–285. [CrossRef]
58. Black, J.S.; Stern, P.C.; Elworth, J.T. Personal and contextual influences on household energy adaptations. *J. Appl. Psychol.* **1985**, *70*, 3–21. [CrossRef]
59. De Groot, J.I.M.; Steg, L. Morality and prosocial behaviour: The role of awareness, responsibility and norms in the norm activation model. *J. Soc. Psychol.* **2009**, *149*, 425–449. [CrossRef] [PubMed]
60. Deci, E.L.; Ryan, R.M. *Intrinsic Motivation and Self-Determination in Human Behaviour*; Plenum Publishing Co.: New York, NY, USA, 1985.
61. Webb, D.; Soutar, G.N.; Mazzarol, T.; Saldaris, P. Self-determination theory and consumer behavioural change: Evidence from a household energy-saving behaviour study. *J. Environ. Psychol.* **2013**, *35*, 59–66. [CrossRef]
62. Miller, N.E.; Dollard, J. *Social Learning and Imitation*; Kegan Paul: London, UK, 1945.
63. Bamberg, S. Applying the stage model of self-regulated behavioral change in a car use reduction intervention. *J. Environ. Psychol.* **2013**, *33*, 68–75. [CrossRef]
64. Bamberg, S.; Hyllenius, P.; Haustein, S.; Welsch, J.; Schreffler, E.; Carreno, M.; Rye, T.; D'Arcier, B.; Zoubir, A. *WPB Final Report—Behaviour Change Models and Prospective Assessment. MAX-Success: Successful Travel Awareness Campaigns and Mobility Management Strategies*; Final report of Work Package Behaviour prepared for the European Commission; European Commission: Brussels, Belgium, 2009.
65. Triandis, H.C. *Interpersonal Behaviour*; Brooks/Cole Publishing Company: Monterrey, CA, USA, 1977.
66. Ajzen, I.; Fishbein, M. *Understanding Attitudes and Predicting Social Behavior*; Prentice-Hall: Englewood Cliffs, NJ, USA, 1980.
67. Bang, H.; Ellinger, A.E.; Hadjimarcou, J.; Traichal, P.A. Consumer Concern, Knowledge, Belief, and Attitude toward Renewable Energy: An Application of the Reasoned Action Theory. *Psychol. Mark.* **2000**, *17*, 449–468. [CrossRef]
68. Bamberg, S. Effects of implementation intentions on the actual performance of new environmentally friendly behaviours-results of two field experiments. *J. Environ. Psychol.* **2002**, *22*, 299–411. [CrossRef]
69. Harland, P.; Staats, H.; Wilke, H.A.M. Explaining proenvironmental intention and behavior by personal norms and the theory of planned behavior. *J. Appl. Soc. Psychol.* **1999**, *29*, 2505–2528. [CrossRef]
70. Staats, H. *Understanding Pro-Environmental Attitudes and Behaviour: An Analysis and Review of Research Based on the Theory of Planned Behaviour*; Ashgate Publishing: Aldershot, UK, 2003.
71. Shove, E.; Southerton, D. Defrosting the Freezer: From Novelty to Convenience: A Narrative of Normalization. *J. Mater. Cult.* **2000**, *5*, 301–319. [CrossRef]
72. Steg, L.; Dreijerink, L.; Abrahamse, W. Factors influencing the acceptability of energy policies: Testing VBN theory. *J. Environ. Psychol.* **2005**, *25*, 415–425. [CrossRef]
73. Davis, F.D. Perceived usefulness, perceived ease of use, and user acceptance of information technology. *MIS Q.* **1989**, *13*, 319–340. [CrossRef]
74. Venkatesh, V.; Davis, F.D. A theoretical extension of the technology acceptance model: Four longitudinal field studies. *Manag. Sci.* **2000**, *46*, 186–204. [CrossRef]
75. Ozag, D.; Jurkiewicz, J. Information System Response Model. In *Human Resources and Their Development—Volume 2*; Marquardt, M.J., Ed.; UNESCO: Paris, France, 2009; Volume 2.

76. Gärling, T.; Eek, D.; Loukopoulos, P.; Fujii, S.; Johansson-Stenman, O.; Kitamura, R.; Pendyala, R.; Vilhelmson, B. A conceptual analysis of the impact of travel demand management on private car use. *Transp. Policy* **2002**, *9*, 59–70. [CrossRef]
77. Steg, L.; Vlek, C. Encouraging pro-environmental behaviour: An integrative review and research agenda. *J. Environ. Psychol.* **2009**, *29*, 309–317. [CrossRef]
78. Lundvall, B.A. *National Systems of Innovation—Towards a Theory of Innovation and Interactive Learning*; Pinter Publishers LTD: London, UK, 1992.
79. Edquist, C. *Systems of Innovation: Technologies, Institutions and Organisations*; Printer Publisher LTD: London, UK, 1997.
80. Nelson, R.R.; Winter, S.G. Toward an Evolutionary Theory of Economic Capabilities. *Am. Econ. Rev.* **1973**, *68*, 440–449.
81. Owen, A.L.; Videras, J. Civic cooperation, pro-environment attitudes, and behavioral intentions. *Ecol. Econ.* **2006**, *58*, 814–829. [CrossRef]
82. Ek, K.; Soderholm, P. The devil is in the details: Household electricity saving behavior and the role of information. *Energy Policy* **2010**, *38*, 1578–1587. [CrossRef]
83. Pedersen, L.H. The dynamics of green consumption: A matter of visibility? *J. Environ. Policy Plan.* **2000**, *2*, 193–210. [CrossRef]
84. Ozaki, R.; Sevastyanova, K. Going hybrid: An analysis of consumer purchase motivations. *Energy Policy* **2011**, *39*, 2217–2227. [CrossRef]
85. Jager, W. Stimulating the diffusion of photovoltaic systems: A behavioural perspective. *Energy Policy* **2006**, *34*, 1935–1943. [CrossRef]
86. Adaman, F.; Karali, N.; Kumbaroglu, G.; Or, I.; Okaynak, B.; Zenginobuz, U. What determines urban households' willingness to pay for CO_2 emission reductions in Turkey: A contingent valuation survey. *Energy Policy* **2011**, *39*, 689–698. [CrossRef]
87. Costanzo, M.; Archer, D.; Aronson, E.; Pettigrew, T. Energy conservation behavior: The difficult path from information to action. *Am. Psychol.* **1986**, *41*, 521–528. [CrossRef]
88. Stern, P. Psychological dimensions of global environmental change. *Annu. Rev. Psychol.* **1992**, *43*, 269–302. [CrossRef]
89. Lindenberg, S. Social rationality versus rational egoism. In *Handbook of Sociological Theory*; Turner, J.H., Ed.; Kluwer Academic/Plenum Publishers: New York, NY, USA, 2001; pp. 635–668.
90. Stern, P.C. What psychology knows about energy conservation. *Am. Psychol. Assoc.* **1992**, *47*, 1224–1232. [CrossRef]
91. Brandon, G.; Lewis, A. Reducing household energy consumption: A qualitative and quantitative field study. *J. Environ. Psychol.* **1999**, *19*, 75–85. [CrossRef]
92. Geller, E.S.; Winett, R.A.; Everett, P.B. *Preserving the Environment. Strategies for Behavioral Change*; Pergamon Press: New York, NY, USA, 1982.
93. Midden, C.J.; Meter, J.E.; Weenig, M.H.; Zievering, H.J. Using feedback, reinforcement and information to reduce energy consumption in households: A field experiment. *J. Econ. Psychol.* **1983**, *3*, 65–86. [CrossRef]
94. Weenig, W.H.; Schmidt, T.; Midden, C.J.H. Social dimensions of neighborhoods and the effectiveness of information programs. *Environ. Behav.* **1990**, *22*, 27–54. [CrossRef]
95. Han, Q.; Nieuwenhijsen, I.; de Vries, B.; Blokhuis, E.; Schaefer, W. Intervention strategy to stimulate energy saving behavior of local residents. *Energy Policy* **2013**, *52*, 706–715. [CrossRef]
96. Abrahamse, W.; Steg, L. Factors related to household energy use and intention to reduce it: The role of psychological and socio-demographic variables. *Hum. Ecol. Rev.* **2011**, *18*, 30–40.
97. Botetzagias, I.; Malesios, C.; Poulou, D. Electricity curtailment behaviors in Greek households: Different behaviors, different predictors. *Energy Policy* **2014**, *69*, 415–424. [CrossRef]
98. De Young, R. New ways to promote proenvironmental behavior: Expanding and evaluating motives for environmentally responsible behavior. *J. Soc. Issues* **2000**, *56*, 509–526. [CrossRef]
99. Gardner, G.T.; Stern, P.C. *Environmental Problems and Human Behavior*; Allyn and Bacon: Boston, MA, USA, 1996.
100. Olsen, M.E. Public acceptance of consumer energy conservation strategies. *J. Econ. Psychol.* **1983**, *4*, 183–196. [CrossRef]

101. Faiers, A.; Cook, M.; Neame, C. Towards a contemporary approach for understanding consumer behavior in the context of domestic energy use. *Energy Policy* **2007**, *35*, 4381–4390. [CrossRef]
102. Gadenne, D.; Sharma, B.; Kerr, D.; Smith, T. The influence of consumers' environmental beliefs and attitudes on energy saving behaviors. *Energy Policy* **2011**, *39*, 7684–7694. [CrossRef]
103. Poortinga, W.; Steg, L.; Vlek, C. Values, environmental concern, and environmental behavior. *Environ. Behav.* **2004**, *36*, 70–93. [CrossRef]
104. Maréchal, K. Not irrational but habitual: The importance of behavioural lock-in in energy consumption. *Ecol. Econ.* **2010**, *69*, 1104–1114. [CrossRef]
105. Abrahamse, W.; Steg, L. How do socio-demographic and psychological factors relate to households' direct and indirect energy use and savings? *J. Econ. Psychol.* **2009**, *30*, 711–720. [CrossRef]
106. Palm, J.; Tengvard, M. Motives for and barriers to household adoption of small-scale production of electricity: Examples form Sweden. *Sustain. Sci. Pract. Policy* **2011**, *7*, 6–15. [CrossRef]
107. Gifford, R. Environmental Psychology Matters. *Annu. Rev. Psychol.* **2014**, *65*, 541–579. [CrossRef]
108. Hondo, H.; Baba, K. Socio-psychological impacts of the introduction of energy technologies: Change in environmental behavior of households with photovoltaic systems. *Appl. Energy* **2010**, *87*, 229–235. [CrossRef]
109. Brosch, T.; Patel, M.K.; Sander, D. Affective influences on energy-related decisions and behaviors. *Front. Energy Res.* **2014**, *2*, 11. [CrossRef]
110. Samuelsom, C.; Biek, M. Attitudes toward energy conservation: A confirmatory factor analysis. *J. Appl. Soc. Psychol.* **1991**, *21*, 549–569. [CrossRef]
111. Wang, Z.; Zhanga, B.; Yinc, J.; Zhanga, Y. Determinants and policy implications for household electricity-saving behavior: Evidence from Beijing, China. *Energy Policy* **2011**, *39*, 3550–3557. [CrossRef]
112. Sardianou, E. Estimating energy conservation patterns of Greek households. *Energy Build.* **2008**, *40*, 1084–1093. [CrossRef]
113. Gram-Hansen, K. Boligers energiforbrug—Sociale og tekniske forklaringer pa forskelle. *By Og Byg Result.* **2003**. Available online: https://www.google.com/url?sa=t&rct=j&q=&esrc=s&source=web&cd=2&ved=2ahUKEwjoo-uU2dflAhUCE4gKHVtyCXUQFjABegQIBBAC&url=https%3A%2F%2Fwww.osti.gov%2Fetdeweb%2Fservlets%2Fpurl%2F20405565&usg=AOvVaw3ZhUYA4gjUI_50k7t2Gl6Y (accessed on 6 November 2019).
114. Carlsson-Kanyama, A.; Linden, A.L.; Ericsson, B. Residential energy behaviour: Does generation matter? *Int. J. Consum. Stud.* **2005**, *29*, 239–252. [CrossRef]
115. Carlsson-Kanyamaa, A.; Lindén, A.L. Energy efficiency in residences-Challenges for women and men in the North. *Energy Policy* **2007**, *35*, 2163–2172. [CrossRef]
116. Vringer, C.R. *Analysis od the Energy Requirement for Household Consumption*; Utrecht University: Utrecht, The Netherlands, 2005.
117. Whitmarsh, L. Behavioural responses to climate change: Asymmetry of intentions and impacts. *J. Environ. Psychol.* **2009**, *29*, 12–23. [CrossRef]
118. Jansson, J.; Marell, A. Green consumer behaviour: Determinants of curtailment and eco-innovation adoption. *J. Consum. Mark.* **2010**, *27*, 358–370. [CrossRef]
119. Fischer, C. Feedback on household electricity consumption: A tool for saving energy? *Energy Effic.* **2008**, *1*, 79–104. [CrossRef]
120. Guerin, D.A.; Yust, B.L.; Coopet, J.G. Occupant predictors of household energy behaviour and consumption change as found in energy studies since 1975. *Fam. Consum. Sci. Res. J.* **2000**, *29*, 48–80. [CrossRef]
121. Thøgersen, J. How May Consumer Policy Empower Consumers for Sustainable Lifestyles? *J. Consum. Policy* **2005**, *28*, 143–177. [CrossRef]
122. Ritchie, J.R.B.; McDougall, G.H.G.; Claxton, J.D. Complexities of Household Energy Consumption and Conservation. *J. Consum. Res.* **1981**, *8*, 233–242. [CrossRef]
123. Wilson, C.; Dowlatabadi, H. Models of decision making and residential energy use. *Annu. Rev. Environ. Resour.* **2007**, *32*, 169–203. [CrossRef]
124. Gardner, G.T.; Stern, P.C. The short list: Most effective actions U.S. households can take to limit climate change. *Environment* **2008**, *50*, 12–25. [CrossRef]
125. Dietz, T.; Gardner, G.T.; Gilligan, J.; Stern, P.C.; Vandenbergh, M.P. Household actions can provide a behavioral wedge to rapidly reduce US carbon emissions. *Proc. Natl. Acad. Sci. USA* **2009**, *106*, 18452–18456. [CrossRef]

126. Stern, P. Information, incentives, and proenvironmental consumer behavior. *J. Consum. Policy* **1999**, *22*, 461–478. [CrossRef]
127. Van Houwelingen, J.; Van Raaij, W. The effect of goal-setting and daily electronic feedback on in-home energy use. *J. Consum. Res.* **1989**, *16*, 98–105. [CrossRef]
128. Boardman, B. New directions for household energy efficiency: Evidence from the UK. *Energy Policy* **2004**, *32*, 1921–1933. [CrossRef]
129. Linden, A.L.; Carlsson-Kanyama, A.; Eriksson, B. Efficient and inefficient aspects of residential energy behavior: What are the policy instruments for change? *Energy Policy* **2006**, *34*, 1918–1927. [CrossRef]
130. Dwyer, W.O.; Leeming, F.C.; Cobern, M.K.; Porter, B.E.; Jackson, J.M. Critical Review of Behavioral Interventions to Preserve the Environment—Research Since 1980. *Environ. Behav.* **1993**, *25*, 275–321. [CrossRef]
131. Georg, S. The social shaping of household consumption. *Ecol. Econ.* **1999**, *28*, 455–466. [CrossRef]
132. Cook, S.W.; Berrenberg, J.L. Approaches to encouraging conservation behavior: A review and conceptual framework. *J. Soc. Issues* **1981**, *37*, 73–107. [CrossRef]
133. Heiskanen, E.; Johnson, M.; Robinson, S.; Vadovics, E.; Saastamoinen, M. Low-carbon communities as a context for individual behavioural change. *Energy Policy* **2010**, *38*, 7586–7595. [CrossRef]
134. Schultz, P.W.; Nolan, J.; Cialdini, R.; Goldstein, N.; Griskevicius, V. The constructive, destructive, and reconstructive power of social norms. *Psychol. Sci.* **2007**, *18*, 429–434. [CrossRef]
135. Bamberg, S.; Möser, G. Twenty years after Hines, Hungerford, and Tomera: A new meta-analysis of psycho-social determinants of pro-environmental behaviour. *J. Environ. Psychol.* **2007**, *27*, 14–25. [CrossRef]
136. Middlemiss, L. Influencing Individual Sustainability: A Review of the Evidence on the Role of Community-Based Organisations. *Int. J. Environ. Sustain. Dev.* **2008**, *7*, 78–93. [CrossRef]
137. Gatersleben, B.; Steg, L.; Vlek, C. Measurement and determinants of environmentally significant consumer behavior. *Environ. Behav.* **2002**, *34*, 335–362. [CrossRef]
138. Eriksson, L.; Garvill, J.; Nordlund, A.M. Interrupting habitual car use: The importance of car habit strength and moral motivation for personal car use reduction. *Transp. Res.* **2008**, *11*, 10–23. [CrossRef]
139. Nordlund, A.M.; Garvill, J. Effects of values, problem awareness, and personal norm on willingness to reduce personal car use. *J. Environ. Psychol.* **2003**, *23*, 339–347. [CrossRef]
140. Lee, E.; Park, N.; Han, J.H. Gender Difference in Environmental Attitude and Behaviors in Adoption of Energy-Efficient Lighting at Home. *J. Sustain. Dev.* **2013**, *6*, 36–50. [CrossRef]
141. Stern, P.C.; Gardner, G. Psychological research and energy policy. *Am. Psychol.* **1981**, *36*, 329–342. [CrossRef]
142. Schultz, P.W. Strategies for promoting proenvironmental behavior: Lots of tools but few instructions. *Eur. Psychol.* **2014**, *19*, 107. [CrossRef]
143. Ehrhardt-Martinez, K.; Donnelly, K.A.; Laitner, J.A. *Advanced Metering Initiatives and Residential Feedback Programs: A Meta-Review for Household Electricity-Saving Opportunities*; American Council for an Energy-Efficient Economy: Washington, DC, USA, 2010.
144. Asare-Bediako, B.; Ribeiro, P.F.; Kling, W.L. Integrated energy optimization with smart home energy management systems. In Proceedings of the Paper presented at the 3rd IEEE PES International Conference and Exhibition on Innovative Smart Grid Technologies (ISGT Europe), Berlin, Germany, 14–17 October 2012.
145. Steg, L.; Perlaviciute, G.; van der Werff, E. Understanding the human dimensions of a sustainable energy transition. *Front. Psychol.* **2015**, *6*, 805. [CrossRef]
146. Vasseur, V.; Kemp, R. The adoption of PV in the Netherlands: A statistical analysis of adoption factors. *Renew. Sustain. Energy Rev.* **2015**, *41*, 483–494. [CrossRef]
147. Vasseur, V. *A Sunny Future for Photovoltaic Systems in the Netherlands? An Analysis of the Role of Government and Users in the Diffusion of an Emerging Technology*; Maastricht University: Maastricht, The Netherlands, 2014.

Article

Modeling Future Energy Demand and CO_2 Emissions of Passenger Cars in Indonesia at the Provincial Level

Qodri Febrilian Erahman, Nadhilah Reyseliani, Widodo Wahyu Purwanto * and Mahmud Sudibandriyo

Chemical Engineering Department, Faculty of Engineering, Universitas Indonesia, Depok 16424, Indonesia
* Correspondence: widodo@che.ui.ac.id

Received: 9 July 2019; Accepted: 12 August 2019; Published: 17 August 2019

Abstract: The high energy demand and CO_2 emissions in the road transport sector in Indonesia are mainly caused by the use of passenger cars. This situation is predicted to continue due to the increase in car ownership. Scenarios are arranged to examine the potential reductions in energy demand and CO_2 emissions in comparison with the business as usual (BAU) condition between 2016 and 2050 by controlling car intensity (fuel economy) and activity (vehicle-km). The intensity is controlled through the introduction of new car technologies, while the activity is controlled through the enactment of fuel taxes. This study aims to analyze the energy demand and CO_2 emissions of passenger cars in Indonesia not only for a period in the past (2010–2015) but also based on projections through to 2050, by employing a provincially disaggregated bottom-up model. The provincially disaggregated model shows more accurate estimations for passenger car energy demands. The results suggest that energy demand and CO_2 emissions in 2050 will be 50 million liter gasoline equivalent (LGE) and 110 million tons of CO_2, respectively. The five provinces with the highest CO_2 emissions in 2050 are projected to be West Java, Banten, East Java, Central Java, and South Sulawesi. The projected analysis for 2050 shows that new car technology and fuel tax scenarios can reduce energy demand from the BAU condition by 7.72% and 3.18% and CO_2 emissions by 15.96% and 3.18%, respectively.

Keywords: energy demand; CO_2 emissions; Indonesia

---

1. Introduction

Since 2013, the transport sector has consumed more energy than any other sector in Indonesia. Approximately 40% of the energy demand (260.1 million BOE) in Indonesia is attributable to the transport sector [1], with road transport being the largest contributor. This situation is predicted to increase, due to the growth of car ownership.

Transportation plays an important role in modern society in terms of supporting the mobility of people; however, it also creates a major problem for the environment. CO_2 emissions in the road transport sector are mostly contributed by the use of passenger cars. This situation is worsened by the lack of improvements to the land transportation system. To ensure mobility under the present circumstances, most people choose to own a private car. The growth in car ownership is considered to be mainly responsible for rising energy demand. Passenger cars in Indonesia mostly consume gasoline, and high demand for gasoline has resulted in Indonesia's dependence on imported petroleum products [2]. Car ownership has a strong correlation with GDP per capita, as shown in many previous studies, including Dargay and Gately [3], Dargay and Gately [4], Dargay and Gately [5], Dargay, Gately and Sommer [6], Leaver, Samuelson and Leaver [7], and Wu, Zhao and Ou [8]. These studies suggest that the GDP per capita can affect the level of energy demand.

The issuing of Presidential Decrees 61/2011 and 71/2011 [9,10] mandated a mitigation plan for greenhouse gas emissions for each province. Based on these regulations, provincial governments were asked to prepare action plans to the reduction of CO_2 emissions. The action plans can be carried out by controlling the intensity and activity of passenger cars. The intensity is related to car technology, while car activity is related to car utilization. Certain policies for controlling the intensity and activity of passenger cars should be encouraged in order to decrease energy demand and CO_2 emissions [11,12]. Therefore, the historical energy demand from the use of passenger cars in each province should be known.

Previous studies have shown that transport energy demand can be projected through top-down models (e.g., Zhang et al. [13], Lu et al. [14], and Chai et al. [15]); however, to determine the impact of technological change, the energy demand projection for the road transport sector should be conducted using a bottom-up model [16]. Other studies have implemented a bottom-up model for projecting the transport energy demand (e.g., Eom and Schipper [17], Ma et al. [18], Baptista et al. [19], Ko et al. [20], and Deendarlianto et al. [21]). However, these studies have mostly been conducted at country level, whereas, because disparities exist among regions, this study was conducted at the provincial level. Moreover, the study contributes to estimating the passenger car energy demand by modeling the technological changes and the activities of the passenger car and to find out which is the best policy for lowering the energy demand and CO_2 emissions. This paper aims to model the future energy demand and CO_2 emissions of passenger cars in Indonesia by province in past (2010–2015) and future (2016–2050) periods.

The remainder of this paper is structured as follows. Section 2 proposes the methodology. Section 3 presents the results and discussion, and Section 4 provides the conclusions.

2. Methods

This section explains the methodology for assessing future energy demand and CO_2 emissions using a bottom-up model. Figure 1 explains the methodological structure of the current study.

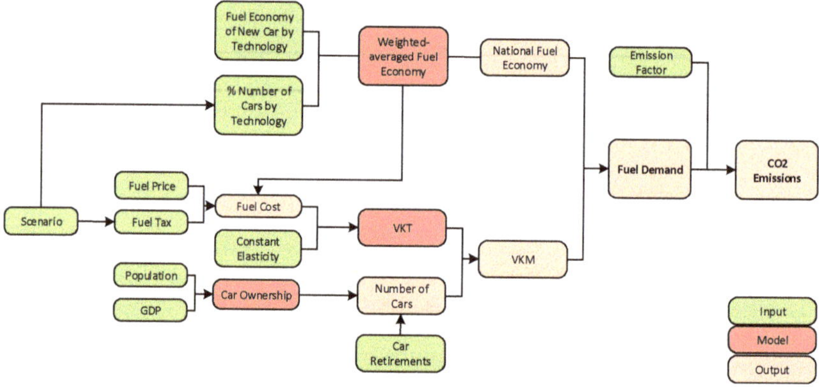

Figure 1. Schematic diagram of the methodology used.

As can be seen in Figure 1, the structures consist of the input, the model and the output. Input includes everything that is to be processed in the model, including data and scenarios. The model consists of car ownership, vehicle kilometers traveled (VKT), and weighted average fuel economy. These aspects of the model will generate the intermediate output of VKM and fuel economy, from which the fuel demand and CO_2 emissions are derived. This structure is applied for each province, and subsequently, the results are aggregated to obtain the national results.

2.1. *Provinces of Indonesia*

Administratively, Indonesia consists of 34 provinces, but the current study analyzed only 33 to adjust to the available data, and also because of the emergence of new provinces in Kalimantan. Each province has its own local government, governor, and legislative body. Spatially, Indonesia can be divided into five major regions: Sumatra, Java, Kalimantan, Sulawesi, and Nusa Tenggara-Maluku-Papua. Table 1 presents the related details.

Table 1. Five regions of Indonesia.

Region	Province	Region	Province
Sumatra	Aceh	Kalimantan	West Kalimantan
	North Sumatra		Central Kalimantan
	West Sumatra		South Kalimantan
	Riau		East Kalimantan
	Jambi	Sulawesi	North Sulawesi
	South Sumatra		Central Sulawesi
	Bengkulu		South Sulawesi
	Lampung		Southeast Sulawesi
	Bangka Belitung Islands		Gorontalo
	Riau Islands		West Sulawesi
Java	DKI Jakarta	Nusa Tenggara–Maluku–Papua	West Nusa Tenggara
	West Java		East Nusa Tenggara
	Central Java		Maluku
	D.I.Y		North Maluku
	East Java		West Papua
	Banten		Papua
	Bali		

Figure 2 shows the profile of Indonesia's territories according to their populations, which are highly concentrated in the west. The capital city of Indonesia known as the Special Capital Region of Jakarta (DKI Jakarta) is located in the Java region, contributing to the fact that this region is the most densely populated. These population trends are expected to continue if the government does not promote greater equity among the provinces.

Figure 2. Spatial population profiles among regions in Indonesia.

2.2. Input Data

Data such as provincial GDP, the number of passenger cars, the size of province area, and population are sourced from the Central Bureau of Statistics of Indonesia [22]. Energy demand for the transport sector, along with fuel price data, were collected from the Ministry of Energy and Mineral Resources of Indonesia [1]. Annual car sales data, which are categorized by engine displacement, were obtained from the Association of Indonesian Automotive Industries (Gaikindo) [23]. The Central Bureau of Statistics of Indonesia provides population projections until 2050 [24], and the projected provincial population takes into account the effect of urbanization. The provincial GDPs are based on commodity prices in the year 2000, and the projections are obtained using GDP growth until 2050 [25]. Finally, the data is inputted into the model.

2.3. Car Ownership

Car ownership exhibits a close relationship with GDP per capita [4]. This empirical relationship follows the Gompertz model, which has been developed in various studies [3–6]. It explains that, over the long term, the relationship between car ownership and GDP per capita corresponds to the following equation:

$$CO_i = CO_i^* \times e^{\alpha_i e^{\beta_i GDPP_i}} \tag{1}$$

where CO is the car ownership (vehicles/1000 people), CO^* is the saturated car ownership, $GDPP$ is GDP per capita, i is the province, and α and β are the constants that determine the shape of the curve. The constants α and β can be obtained according to the following equation [8].

$$ln\left(ln\frac{CO_i^*}{CO_i}\right) = ln(-\alpha_i) + \beta_i \cdot GDPP_i \tag{2}$$

In the equation, α and β are constants to determine the curve shape. The relationship between GDP per capita and long-term car ownership forms an S-shaped curve. This S-shape implies that at a relatively low level of GDP per capita, the growth rate of car ownership will rise slowly, then will grow dramatically at a certain GDP per capita level, and will finally slow down again at a high level of GDP per capita until reaching a steady state, which is known as car ownership saturation [5].

The car ownership saturation is a condition in which GDP per capita continues to increase, while car ownership remains unchanged. Previous studies have suggested that there is a relationship between population density and the saturation level of car ownership [7]. For example, Leaver established a relationship between population density and car ownership saturation [7]. The higher the population density, the faster car ownership saturation occurs, and the current study uses this finding to determine the saturation level of car ownership for each province, as shown in the following equation:

$$CO^* = 606.5e^{(0.007 \times D)} \tag{3}$$

where D is population density. Since the analysis is conducted at the provincial level, the effects of urbanization have been included in the projected population data. Figure 3 summarizes the scheme of the car ownership projection model.

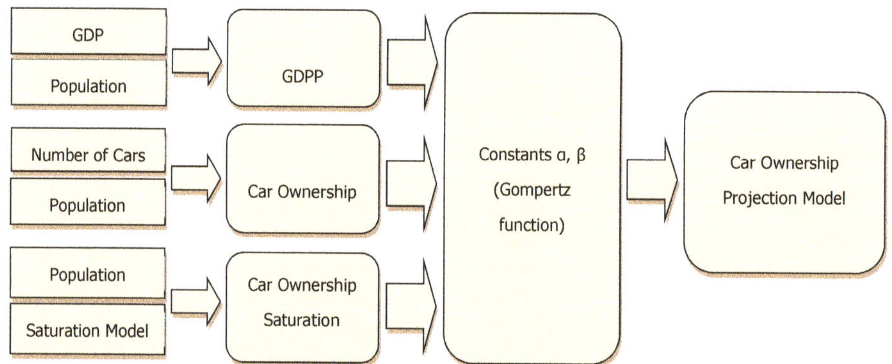

Figure 3. Scheme of the car ownership projection model.

2.4. Car Fuel Economy

Fuel economy is reported in units of L/100 km. National fuel economy is calculated from the weighted average of new and existing car shares and their respective fuel economies. The fuel economy of new cars is taken from a weighted average of annual car sales by fuel type, i.e., gasoline vs. diesel cars. Fuel economy is further characterized according to engine size: 800 < cc < 1200, 1200 < cc < 1500, 1500 < cc < 3000, and 3000 < cc for gasoline cars; and 1500 < cc < 3000 for diesel cars. Cars with an engine size of 800 < cc < 1200 are referred to as low-cost green cars (LCGC) [26].

In the projected scenario, due to the presence of new car technology (e.g., plug-in hybrid [PHEV] and electric vehicle [EV] technology), the fuel economy of a new car is weighted by the share of each type of car—gasoline, diesel, PHEV, and EV—according to the following equations.

$$FE_{NC} = \sum_{j} FE_j \times \%C_j. \tag{4}$$

$$FE = FE_{NC} \times \%C_{NC} + FE_{RC} \times \%C_{RC} \tag{5}$$

where FE is fuel economy, $\%C$ is the percentage of cars, and j is the type of car based on its technology (e.g., gasoline, diesel, PHEV, or EV). NC is new car and RC is the rest of the cars. Figure 4 describes the fuel economy aggregation scheme based on car technology.

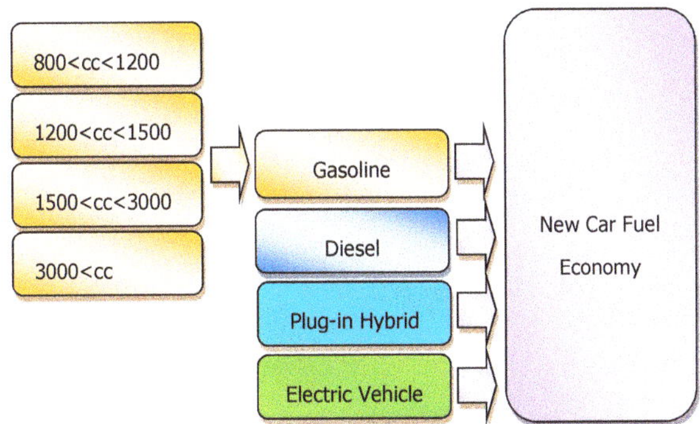

Figure 4. Fuel economy aggregation scheme based on car technology.

The historical fuel economy (2010–2015) for an engine size of 800 < cc < 1200, which is in the LCGC category, is 5.0 L/100 km [27]. Cars with engine sizes of 1200 < cc < 1500, 1500 < cc < 3000, and 3000 < cc have the highest market share and fuel economies of 8.20 L/100 km, 10.10 L/100 km, and 12.40 L/100 km, respectively [28–30]. Diesel cars, which have a fuel economy of 6.97 L/100 km [12], are considered to be 20% more efficient than gasoline cars. Car fuel economy for engine sizes 1200 < cc < 1500 and 1500 < cc < 3000 was contributed by sedan and MPV (Multi-Purpose Vehicle) types of vehicle, while for cars with engine size 3000 < cc, this was contributed by Sedan and SUV (Sport Utility Vehicle) types. The percentages of sedans, MPVs and SUVs are 6.1%, 93.2, and 0.6% of total cars, respectively.

The fuel economy for PHEV and EV cars was not applied in the historical situation, since their market share was zero until 2015. Figure 5 describes the aggregation scheme of the weighted average of fuel economy between new and other cars.

Fuel economy for new cars is considered starting in 2010; for the remainder of the cars, fuel economy before 2010 is assumed based on the IEA report [31].

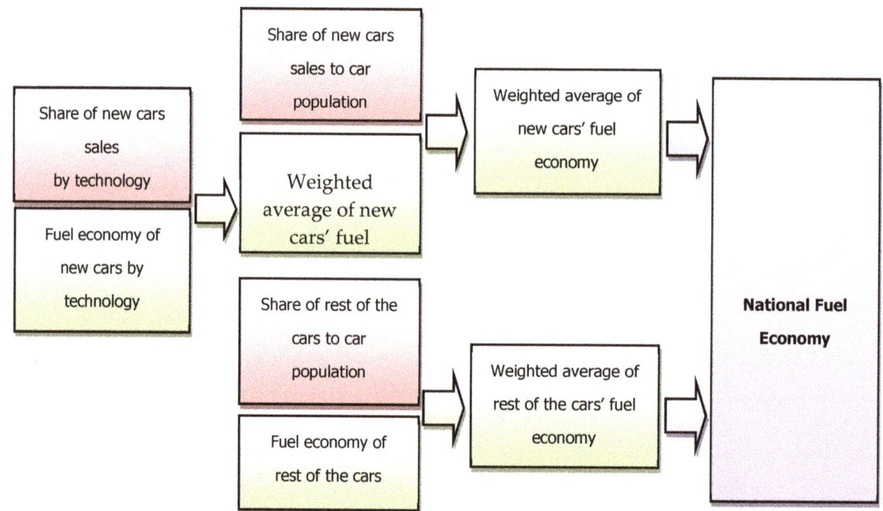

Figure 5. Aggregation of fuel economy between new and other cars.

2.5. Vehicle Kilometers Traveled

Vehicle kilometers traveled, VKT, is defined as the annual kilometers traveled for a single car. Previous studies show an inverse relationship between VKT and fuel price, meaning that car users will tend to reduce unnecessary travel when the fuel price increases. The extent to which VKT varies with changing fuel price can be modeled by the value of elasticities, according to the following equation [32].

$$VKT_i = VKT_i' \times \left(\frac{FC_i}{FC_i'}\right)^{\varepsilon} \tag{6}$$

where $VKT_.$ represents the vehicle kilometers traveled in a given year, $VKT'_.$ is the vehicle kilometers traveled in the previous year, $FC_.$ is the fuel cost in a given year, FC' is the fuel cost in the previous year, and ε is the elasticity. VKT data per province can be obtained through calculations of fuel consumption, fuel economy, and number of vehicles in the historical year (2012–2015). Previous studies described that annual car travel is also influenced by car fuel economy [33]; therefore, the current study prefers to use fuel cost instead of fuel price in order to more effectively assess the impact of real situations on the behavior of private car users. Fuel cost is described as the retail fuel price multiplied by the national fuel economy. In the projection, the retail fuel price is obtained by the summation of crude oil price, refinery margin, and distribution fees to customers, and fuel taxes. Crude oil price is based on the US Energy Information Administration outlook [34], and the refinery margin follows the Asia refining margin outlook [35]. Meanwhile, the distribution cost is assumed to remain constant [36]. The sum of total cars traveling in a certain year is defined as car activity, VKM.

2.6. Energy Demand and CO_2 Emissions

Energy demand is defined in units of liter gasoline equivalent (LGE). Cars that consume other fuels, such as diesel oil, should be converted into LGE using heating value comparisons between gasoline and diesel oil, where the heat value for diesel, biodiesel and gasoline is 35,327, 36,131 and 31,795 kJ/L, respectively. Energy demand can then be calculated according to the following equation:

$$E_i = VKM_i \times FE. \tag{7}$$

where E is the energy demand, and VKM represents vehicle kilometers, which represents the total number of cars traveling annually. Once the energy demand is determined, then CO_2 emissions can be calculated using the following equation:

$$G_i = E_i \times EF_k \tag{8}$$

where G represents the CO_2 emissions, EF is the emission factor, and k is the type of fuel (e.g., gasoline, diesel oil, and electricity). Equations (7) and (8) are consistent with the ASIF equation, which is widely used for calculating CO_2 emissions. Emission factors were obtained from the Ministry of Environment of Indonesia, which in turn based them on information from the Intergovernmental Panel on Climate Change (IPCC). Therefore, the emission factors for gasoline, diesel, biodiesel B100, and electricity were 69.3, 74.1, 62.9, and 224.4 kg CO_2/GJ, respectively [37,38]. Moreover, the electricity emission factor was based on the weighted-average data from all kinds of power plants in Indonesia [38].

2.7. Model Validation

The results of the analysis need to be validated to determine the accuracy of the model. This is accomplished by comparing the results with the fuel demand in 2010–2015 using the standard error of the estimate. The standard error of the estimate is a measure of the accuracy of predictions. It indicates how far data points are from the prediction line of the average. The following is the equation of the standard error of the estimate.

$$\sigma_{est} = \sqrt{\frac{\sum(E - E')^2}{N}}$$

where σ_{est} is the standard error of the estimate, E denotes the data points, E' is the predicted value, and N is the number of data points.

2.8. Scenarios

Scenarios for reducing CO_2 emissions from car utilization can be developed by managing the intensity and activity of cars. Controlling the intensity of cars can be achieved by encouraging the uptake of new technologies that allow for better fuel economy and emissions reduction. Therefore, the market share of new cars with better fuel technology should be increased in order to improve fuel economy. To purchase the most efficient cars in the market, consumers must first understand the efficiency features of the cars under consideration [39]. Therefore, fuel economy labeling should become a required policy to support the introduction of new car technologies that enable better fuel economy. Fuel economy labeling is carried out by obligating car manufacturers or dealers to provide information on the fuel economy of new cars. Car labeling policies are also useful as an important basis for other policies, such as fuel economy standards [12].

Car activity can be managed by regulating the fuel price, so that car users will limit unnecessary travel. The policy required to support this scenario is fuel taxes arrangement [12]. Fuel taxes are an appropriate policy for reducing car travel, because the higher the fuel prices are, the more people will reduce car travel, especially for unnecessary trips. Fuel taxes can provide significant incremental incentives to save fuel and can be integral to any policy package to promote sustainable transport, whereas fuel subsidies are considered to be counterproductive [12]. Fuel taxes also provide revenues to pay for infrastructure costs and to develop sustainable transport. Therefore, scenarios exploring these various policies are created in the current study and are divided into three parts: BAU, new car technology, and fuel tax regulation. These scenarios are intended for use in the projections from 2016 to 2050.

a. Business as Usual Scenario (BAU)

This scenario assumes that the available car technology is limited to gasoline and diesel cars; however, new car fuel economy is expected to improve. Projections for technological developments related to new car fuel economy follow recent developments in non-OECD countries for fuel economy improvement rates [31]. Fuel economy improvement can be applied for gasoline and diesel cars until 2050. The share of cars based on technology follows the historical pattern (2015), in which the shares of car sales for gasoline and diesel cars are 83% and 17%, respectively. For PHEV and EV, on the other hand, the sales remain at zero due to the lack of government initiatives encouraging sales. In the BAU scenario, the fuel tax percentage follows the current situation, which is 15% of the fuel price, and it is assumed that there will be no change in the following years.

b. Car Technology Scenario

The car technology scenario is related to the government's national energy plan for the market penetration of electric vehicles, as stated in Presidential Decree 22/2017 [40]. This scenario assumes that market penetration for PHEV and EV cars is growing significantly. The penetration for PHEV and EV cars follows the IEA's Blue Map scenario [41], wherein to reduce significant global emissions, it is necessary that the 2050 sales mix for PHEV and EV is equal to at least half of total annual car sales [41]. Therefore, the sales mix for PHEV and EV in 2050 is targeted at 50%, while the remaining 50% constitutes mixed sales of diesel and gasoline cars. Table 2 describes the percentage of car sales by type and scenario. The success of car technology scenarios for CO_2 emission reduction hinges on the significant decrease in the electricity emission factor. Based on the Blue Map scenario, the electricity emissions factor should be decreased to almost zero in 2050 [41]; therefore, the electricity emission factor for the car technology scenario is assumed to decrease gradually, reaching 27.8 kg CO_2/GJ in 2050. The target of reducing the emission factor of the electricity can be conducted by increasing the supply of electricity from renewable sources, i.e., geothermal, hydro, solar, wind and biomass.

Table 2. Comparison of percentages of new car sales by car technology and scenario.

Type of Car Technology	BAU Scenario (%)				Car Technology Scenario (%)			
	2020	2030	2040	2050	2020	2030	2040	2050
Regular Gasoline	83	83	83	83	78	68	48	33
Diesel	17	17	17	17	17	17	17	17
Plug-in Hybrid (PH)	0	0	0	0	5	10	25	35
Electric Vehicle (EV)	0	0	0	0	0	5	10	15

c. Fuel Tax Regulation Scenario

This scenario aims to study the effect of car activity on energy demand through the regulation of fuel tax. Changes in fuel cost could affect the VKT, which in turn could affect the VKM. The responses of car users to rising fuel costs are different in each province, and this is indicated through the elasticity. In 2015, the decrease in global crude oil prices caused a decline in fuel prices. The government took advantage of this situation by eliminating fuel subsidies, particularly for the transport sector. Since then, the government has imposed an economic price for gasoline. After the cessation of subsidies, tax policy became recognized as an effective instrument for controlling car travel. Currently, the two kinds of applied fuel tax are value added tax and motor vehicle fuel tax, with values of 10% and 5% of the retail price, respectively. Therefore, the total applied accumulated tax is 15% of the retail price.

A comparison with other countries in the ASEAN region shows that in 2012, the total tax related to fuel demand in these countries ranged from 4–36% [42]. Therefore, to make our scenario more plausible, the fuel tax was set at 30%. The fuel tax scenario assumes no changes in the share of new car sales, and the fuel economy of new cars follows the BAU scenario. Therefore, any changes in energy demand and CO_2 emissions are due solely to changes in car activity. Table 3 summarizes the comparison of assumptions among scenarios.

Table 3. Comparison of assumptions among scenarios.

Scenario		Annual Rate of Fuel Economy Improvement	Target Share of Car Sales to Total Car Sales in 2050	Fuel Tax Rate
Business as Usual (BAU)		Gasoline and Diesel Car 0.09%	No Change	15%
Car Technology	PH/EV	Gasoline, Diesel, PH/EV Car 0.09%; 0.09%; 1.40%	50% of PH/EV	15%
Fuel Taxes	Tax 30%	Gasoline and Diesel Car 0.09%	No Change	30%

3. Results and Discussion

3.1. Historical Results

3.1.1. GDP Per Capita

GDP data were collected from 2000 to 2015. The national GDP is an aggregation of all provincial GDPs. Each province contributes independently to the national GDP, and there are disparities among provinces. Based on provincial GDP data, it can be determined that 57% of the national GDP is from DKI Jakarta, East Java, West Java, and Central Java. However, the prosperity level is more suitably represented by GDP per capita. Table 4 describes the GDP per capita for each province.

Table 4. GDP per capita of provinces, 2000–2015 (Rp).

No.	Province	GDP Per Capita (Rp)				Annual Growth
		2000	2005	2010	2015	
1	Aceh	4,995,043	5,568,355	6,427,395	8,208,249	4.29%
2	North Sumatra	5,936,151	7,136,919	9,112,107	11,435,561	6.18%
3	West Sumatra	5,387,147	6,411,608	7,987,615	10,095,631	5.83%
4	Riau	14,034,330	15,108,162	17,531,364	18,746,113	2.24%
5	Jambi	3,964,314	4,583,988	5,622,244	7,397,381	5.77%
6	South Sumatra	5,988,369	6,917,533	8,535,492	10,555,139	5.08%
7	Bengkulu	3,105,780	3,801,072	4,842,777	6,010,136	6.23%
8	Lampung	3,448,223	4,097,222	5,028,805	6,344,406	5.60%
9	Bangka Belitung Islands	7,168,132	7,949,017	8,709,608	10,456,111	3.06%
10	Riau Islands	18,395,851	22,344,514	24,265,039	28,706,274	3.74%
11	DKI Jakarta	27,160,473	32,812,888	41,037,969	52,793,584	6.29%
12	West Java	5,484,062	6,165,875	7,454,209	9,245,740	4.57%
13	Central Java	3,672,917	4,497,646	5,763,579	7,399,348	6.76%
14	D.I.Y	4,317,566	5,140,272	6,068,938	7,463,150	4.86%
15	East Java	5,842,889	7,110,540	9,111,499	12,144,534	7.19%
16	Banten	6,535,249	7,187,098	8,284,732	9,923,154	3.46%
17	Bali	5,702,601	6,227,553	7,391,742	9,499,575	4.44%
18	West Nusa Tenggara	3,041,105	3,568,679	4,444,685	4,713,600	3.67%
19	East Nusa Tenggara	1,992,050	2,285,129	2,666,020	3,214,568	4.09%
20	West Kalimantan	4,803,628	5,533,075	6,875,073	8,405,443	5.00%
21	Central Kalimantan	5,944,899	6,898,169	8,467,974	10,404,069	5.00%
22	South Kalimantan	6,266,482	7,045,690	8,421,300	10,107,667	4.09%
23	East Kalimantan	12,325,552	14,314,410	18,747,036	32,503,297	10.91%
24	North Sulawesi	5,295,832	5,951,651	8,068,150	10,711,207	6.82%
25	Central Sulawesi	3,977,784	4,940,970	6,660,685	8,922,062	8.29%
26	South Sulawesi	3,506,238	4,526,019	6,352,030	8,623,764	9.73%
27	Southeast Sulawesi	3,170,649	3,960,096	5,194,289	6,794,659	7.62%
28	Gorontalo	1,764,308	2,162,664	2,792,392	3,668,652	7.20%
29	West Sulawesi	1,076,863	3,030,552	4,073,206	5,507,867	27.43%
30	Maluku	2,297,113	2,379,840	2,757,219	3,436,217	3.31%
31	North Maluku	2,394,251	2,453,784	2,909,660	3,526,332	3.15%
32	West Papua	1,238,184	8,227,709	12,232,275	19,351,973	97.53%
33	Papua	6,013,255	6,968,230	8,195,795	8,575,849	2.84%

3.1.2. Car Ownership

Table 5 shows car ownership levels in each province between 2000 and 2015. It shows that the province with the highest car ownership level is DKI Jakarta. Other provinces with substantial car ownership levels are Bali, Central Kalimantan, and Riau.

Table 5. Car ownership in provinces, 2000–2015 (Vehicles/1000 People).

No.	Province	2000	2005	2010	2015
1	Aceh	6	15	21	31
2	North Sumatra	14	18	25	36
3	West Sumatra	6	8	24	43
4	Riau	10	40	80	100
5	Jambi	9	17	30	51
6	South Sumatra	8	21	51	87
7	Bengkulu	7	10	19	27
8	Lampung	6	9	10	20
9	Bangka Belitung Islands	5	8	17	41
10	Riau Islands	10	28	73	93
11	DKI Jakarta	148	196	242	345
12	West Java	9	11	13	21
13	Central Java	6	6	13	25
14	D.I.Y	21	32	72	99
15	East Java	12	20	27	37
16	Banten	2	3	8	12
17	Bali	34	97	134	170
18	West Nusa Tenggara	3	7	23	31
19	East Nusa Tenggara	2	8	29	36
20	West Kalimantan	6	20	65	78
21	Central Kalimantan	3	26	83	101
22	South Kalimantan	11	24	43	58
23	East Kalimantan	15	30	56	71
24	North Sulawesi	11	16	32	65
25	Central Sulawesi	9	35	54	68
26	South Sulawesi	8	15	22	32
27	Southeast Sulawesi	1	4	9	17
28	Gorontalo	0	5	63	85
29	West Sulawesi	35	54	72	99
30	Maluku	16	20	21	27
31	North Maluku	1	1	1	3
32	West Papua	18	29	68	92
33	Papua	5	8	20	28

According to the Gompertz model, in long-term projections, car ownership will form an S-curve. The differences in the S-curve shape in each province will depend on the value of α, β, and the saturation level for car ownership. The values of α and β are strongly influenced by the historical relationship between car ownership and provincial GDP per capita, while the saturation level for car ownership will be different in each province due to differences in population density.

Table 6 shows the results of the car ownership analysis, which pertain to the car ownership model and are based on the historical situation, particularly from 2000 to 2015. The R^2 value shows the accuracy of α and β in the linearized Gompertz model (Equation (2)).

Table 6. Results of car ownership analysis by province using the Gompertz model.

No	Province	Pop. Density (Pop/Ha)	α	β	CO* (Vehicles/1000 People)	R^2
1	Aceh	0.76	−8.2	−0.00000013	603.31	0.877
2	North Sumatra	1.77	−5.0	−0.00000005	599.06	0.997
3	West Sumatra	1.11	−10.0	−0.00000013	601.82	0.943
4	Riau	0.62	−36.7	−0.00000017	603.91	0.915
5	Jambi	0.58	−7.5	−0.00000016	604.06	0.919
6	South Sumatra	0.85	−12.3	−0.00000018	602.93	0.937
7	Bengkulu	0.85	−6.8	−0.00000013	602.92	0.951
8	Lampung	2.14	−6.7	−0.00000011	597.53	0.977
9	Bangka Belitung Islands	0.70	−21.4	−0.00000020	603.59	0.930
10	Riau Islands	1.43	−24.9	−0.00000010	600.57	0.791
11	DKI Jakarta	21.28	−3.5	−0.00000004	523.08	0.973
12	West Java	11.12	−5.7	−0.00000006	561.47	0.974
13	Central Java	9.91	−7.9	−0.00000013	565.96	0.957
14	D.I.Y	10.63	−9.1	−0.00000023	563.25	0.917
15	East Java	7.63	−5.0	−0.00000005	575.06	0.884
16	Banten	10.38	−11.8	−0.00000012	564.55	0.905
17	Bali	6.55	−6.9	−0.00000020	579.55	0.697
18	West Nusa Tenggara	2.18	−18.0	−0.00000040	597.36	0.899
19	East Nusa Tenggara	0.94	−18.3	−0.00000063	602.56	0.804
20	West Kalimantan	0.30	−14.5	−0.00000025	605.25	0.813
21	Central Kalimantan	0.14	−19.1	−0.00000025	605.91	0.799
22	South Kalimantan	0.93	−9.1	−0.00000014	602.62	0.870
23	East Kalimantan	0.17	−4.5	−0.00000003	605.81	0.651
24	North Sulawesi	1.52	−6.8	−0.00000011	600.12	0.969
25	Central Sulawesi	0.41	−5.9	−0.00000012	604.77	0.721
26	South Sulawesi	1.77	−5.2	−0.00000007	599.07	0.867
27	Southeast Sulawesi	0.56	−9.5	−0.00000015	604.14	0.955
28	Gorontalo	0.79	−30.9	−0.00000084	603.17	0.826
29	West Sulawesi	0.64	−3.3	−0.00000011	603.80	0.951
30	Maluku	0.23	−4.2	−0.00000009	605.53	0.850
31	North Maluku	0.30	−10.4	−0.00000018	605.22	0.928
32	West Papua	0.07	−4.0	−0.00000004	606.21	0.825
33	Papua	0.08	−17.0	−0.00000020	606.14	0.949

The α value indicates that the Gompertz curve shifts either to the left or to the right along the x-axis. The lower the value of α, the more the Gompertz curve shifts to the right along the x-axis, and thus, the more distant it gets from a saturated condition. The β value indicates the growth rate of car ownership for certain year ranges. The smaller the β is, the higher is the car ownership growth.

Car ownership saturation shows an asymptotic value, where car ownership is in the steady state. As depicted in Table 6, DKI Jakarta has the lowest car ownership saturation level, due to having the highest population density. Therefore, DKI Jakarta will be the first province that will experience saturation.

3.1.3. National Car Fuel Economy

Figure 6 shows the market shares of gasoline cars sold by engine size during 2010–2015. It shows a decline in the share of cars with engine sizes of cc < 1500 and 1500 < cc < 3000 and an increase in the share of cars with an engine size of 800 < cc < 1200 (LCGC). During 2013–2015, the increase in LCGC accounted for a decrease in the sales of cars with larger engine sizes. Figure 6 also shows the shares for gasoline vs. diesel cars during 2010–2015. The higher level of current diesel car sales is because several car manufacturers have started to offer diesel technology in their vehicles. In contrast, PHEV and EV are still not commercially available in the Indonesian automobile market, and therefore their shares remain at zero.

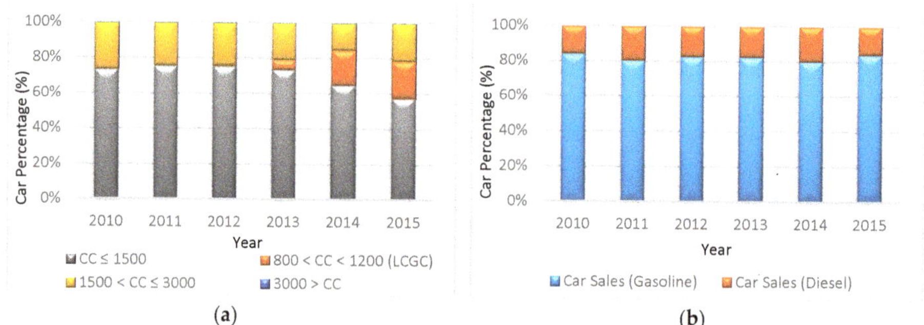

Figure 6. Shares of car sales by (**a**) engine size (gasoline cars) and (**b**) engine type.

Based on market share data, the national car fuel economy showed a decline, as shown in Figure 7. The accumulated car fuel economy describes the average fuel economy for all cars in Indonesia, while the car sales fuel economy describes the fuel economy only for cars that were sold in a given year. Fuel economy for sold cars improved after 2012, which was mainly due to the increasing number of LCGC cars. The fuel economy discrepancy between sold cars and accumulated cars is in the range of 1–1.56 L/100 km, where this discrepancy is estimated to be larger throughout the years.

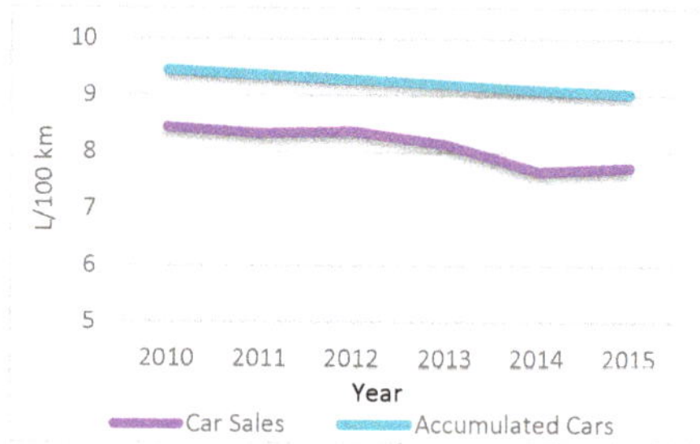

Figure 7. Fuel economy of sold cars and accumulated cars.

3.1.4. Vehicle Kilometers Traveled

Vehicle kilometers traveled, VKT, exhibits disparities between provinces, as seen in Table 7, which indicates the changes in the historical VKT during 2012–2015. VKT changes as fuel cost changes, and the magnitude of thoses changes depends on elasticity.

VKT declines in provinces due to increases in fuel cost. The fuel economy improvement, as shown in Figure 7, is unable to offset the increase in fuel price. Therefore, the total fuel cost is still increasing. Elasticities in the provinces range from −0.067 to −1.051. Elasticity greater than 1 indicates an elastic change in VKT when there is a slight change in fuel cost. An elasticity value less than 1, on the other hand, indicates a small change in VKT with a change in the fuel cost. The East Kalimantan province shows perfect elasticity; therefore, the changes in the fuel cost will be proportional to the VKT changes. Moreover, the highest VKT is observed in Banten. This may be due to Banten's adjacency to the central capital region of DKI Jakarta. Consequently, Banten has many residents who are commuters; these people live in Banten but work in DKI Jakarta.

Table 7. Vehicle Kilometers Traveled in Provinces, 2012–2015 (km/car/year).

No.	Province	2012	2013	2014	2015	Elasticity
1	Aceh	8902	7867	7113	6647	−0.646
2	North Sumatra	11,962	11,319	10,822	10,499	−0.288
3	West Sumatra	15,432	13,916	12,791	12,085	−0.541
4	Riau	9623	9142	8768	8525	−0.268
5	Jambi	5827	5182	4710	4416	−0.613
6	South Sumatra	6799	6110	5600	5281	−0.559
7	Bengkulu	10,242	9574	9062	8733	−0.352
8	Lampung	16,858	15,591	14,629	14,014	−0.409
9	Bangka Belitung Islands	17,743	15,363	13,660	12,621	−0.753
10	Riau Islands	10,563	9909	9406	9082	−0.334
11	DKI Jakarta	4762	4334	4013	3811	−0.492
12	West Java	33,674	31,320	29,523	28,371	−0.379
13	Central Java	10,753	10,106	9608	9286	−0.324
14	D.I.Y	5231	4638	4204	3935	−0.629
15	East Java	10,641	10,024	9547	9239	−0.313
16	Banten	48,432	44,062	40,793	38,728	−0.494
17	Bali	6612	5917	5404	5084	−0.581
18	West Nusa Tenggara	8213	7953	7748	7612	−0.168
19	East Nusa Tenggara	6994	6594	6285	6085	−0.308
20	West Kalimantan	7740	7131	6671	6378	−0.428
21	Central Kalimantan	7889	7311	6872	6590	−0.398
22	South Kalimantan	10,001	9326	8810	8478	−0.365
23	East Kalimantan	10,525	8608	7307	6543	−1.051
24	North Sulawesi	10,696	9600	8791	8284	−0.565
25	Central Sulawesi	5368	4874	4506	4273	−0.504
26	South Sulawesi	18,582	18,108	17,730	17,479	−0.135
27	Southeast Sulawesi	8336	7906	7573	7356	−0.276
28	Gorontalo	9325	8642	8122	7789	−0.398
29	West Sulawesi	3945	3895	3854	3828	−0.067
30	Maluku	11,336	10,035	9085	8497	−0.638
31	North Maluku	38,195	35,925	34,174	33,042	−0.320
32	West Papua	8690	8299	7994	7794	−0.241
33	Papua	17,289	15,564	14,286	13,484	−0.550

3.1.5. Energy Demand and CO_2 Emissions

The energy demand for provinces tends to increase from 2010 to 2015, as depicted in Table 8. The five provinces with the highest energy demand, i.e., West Java, East Java, DKI Jakarta, Central Java, and Riau, are quite similar to the top five provinces in GDP rating. This shows that more than 50% of car energy demand arises from the Java region.

National energy demand is an aggregation of energy demand for all provinces. As depicted in Figure 8, national energy demand increased by 29% from 2010–2015, while GDP increased by 34% for the same period. In other words, energy demand and GDP increased almost proportionally during this time. Although energy demand showed a gradual steady increase, stagnation occurred during 2013–2015. This was caused by the increase in gasoline prices due to government regulation, with the result being that most people reduced unnecessary travel.

The CO_2 emissions profile is quite similar to that of energy demand and shows a gradual increase from 2010 to 2015. About 95% of the total emissions were from gasoline cars, and the remainder were from diesel cars. The emissions from diesel cars resulted from the consumption of a fuel mix of diesel oil and biodiesel that was mandated by the Ministry of Energy and Mineral Resources Regulations 32/2008 and 25/2013 [43,44]. Biodiesel mix usage increased from 1% in 2010 to 10% in 2015. The mandatory biodiesel mix regulation played a role in CO_2 emissions reductions in 2010 and 2015, which were 0.02% and 0.11%, respectively.

Table 8. Car energy demand among provinces, 2010–2015 (LGE).

No	Province	2010	2011	2012	2013	2014	2015
1	Aceh	80,272,962	83,459,252	96,186,659	97,151,788	88,041,546	92,007,033
2	North Sumatera	368,195,076	398,778,700	428,807,615	433,495,984	434,806,571	471,777,332
3	West Sumatera	171,715,475	191,061,362	212,860,361	210,369,069	231,204,316	244,306,713
4	Riau	401,637,066	424,232,500	459,651,842	456,782,973	449,075,161	488,310,136
5	Jambi	51,014,129	57,683,445	65,676,461	71,232,337	65,564,801	68,754,473
6	South Sumatera	243,740,412	285,420,963	309,373,517	349,723,175	318,295,916	335,691,155
7	Bengkulu	30,708,475	32,460,514	37,205,988	39,572,877	37,694,000	40,625,685
8	Lampung	124,065,421	167,065,519	189,579,767	197,602,219	196,189,768	210,202,499
9	Bangka Belitung Islands	35,605,535	37,772,088	62,464,307	62,710,826	62,720,635	64,810,117
10	Kepulauan Riau	122,207,893	129,159,132	139,937,588	141,387,616	138,772,831	149,853,524
11	DKI Jakarta	1,041,349,357	1,128,301,346	1,212,311,139	1,199,831,839	1,153,330,676	1,207,272,570
12	West Java	1,733,899,566	2,105,777,600	2,302,590,272	2,435,271,958	2,371,599,183	2,548,918,144
13	Central Java	427,186,435	563,028,914	626,886,265	658,273,258	666,448,449	720,390,417
14	D.I.Y	121,431,212	128,677,346	139,728,931	133,105,959	124,060,828	129,877,218
15	East Jawa	1,011,926,496	1,069,281,965	1,145,698,694	1,128,630,903	1,102,302,720	1,193,007,164
16	Banten	386,896,273	421,268,116	454,630,752	497,883,116	461,856,866	490,402,758
17	Bali	323,614,972	342,779,339	354,166,716	328,268,247	308,332,155	324,423,028
18	West Nusa Tenggara	81,830,739	86,357,530	90,167,797	92,074,655	93,458,079	102,697,056
19	East Nusa Tenggara	90,600,494	95,731,381	95,951,047	92,317,948	92,448,663	100,106,854
20	West Kalimantan	208,164,352	220,190,058	223,421,205	208,194,952	201,069,751	214,994,820
21	Central Kalimantan	136,805,275	144,669,366	148,016,116	143,807,128	139,959,792	150,126,893
22	South Kalimantan	146,094,739	154,448,564	168,224,184	165,499,922	163,589,675	176,070,648
23	East Kalimantan	194,524,071	206,911,074	222,905,541	193,585,308	169,301,566	169,552,498
24	North Sulawesi	73,766,474	78,123,605	84,540,991	118,237,705	111,633,930	117,658,795
25	Central Sulawesi	71,377,751	75,552,719	77,884,330	72,505,069	70,734,641	75,027,803
26	South Sulawesi	312,293,952	353,372,892	370,622,887	393,444,364	386,676,987	426,360,187
27	Southeast Sulawesi	15,219,392	18,787,111	21,832,002	25,485,292	25,825,223	28,056,494
28	Gorontalo	58,001,180	61,335,857	65,203,277	61,777,834	63,334,762	67,934,623
29	West Sulawesi	31,033,577	35,094,179	36,785,788	39,566,758	39,304,488	43,651,669
30	Maluku	35,038,104	37,132,142	38,651,965	35,763,610	32,921,279	34,434,362
31	North Maluku	4,312,530	4,543,853	6,847,329	8,468,084	9,045,675	9,782,882
32	West Papua	42,763,281	45,158,205	46,431,041	46,741,345	51,636,318	56,308,624
33	Papua	93,191,714	98,682,995	101,732,042	96,536,439	101,630,751	107,288,022
	Indonesia	8,270,484,380	9,282,299,634	10,036,974,419	10,235,300,557	9,962,867,999	10,660,682,198

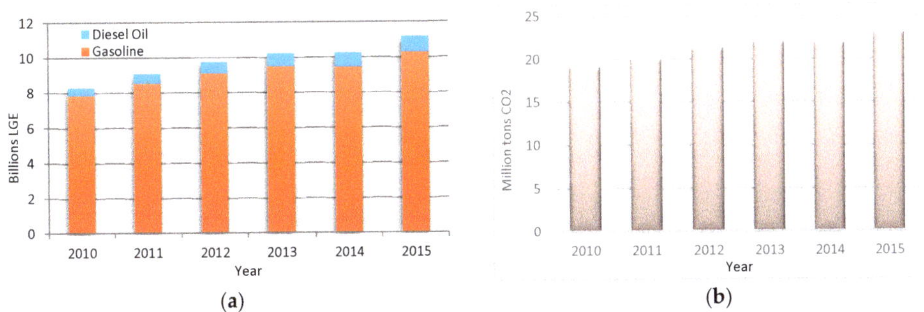

Figure 8. Historical (**a**) energy demand and (**b**) CO_2 emissions, 2010–2015.

However, efforts for reducing CO_2 emissions can be more easily understood through examination of the intensity of CO_2 emissions per car activity. In 2010, the CO_2 emissions intensity per car activity was 207 g CO_2/km, while in 2015 it decreased to 198 g CO_2/km. This indicates a gradual decline of 0.94% per year.

With respect to emissions intensity per car activity, a comparison between countries listed on the International Council on Clean Transportation (ICCT) report in 2010 showed the following: in Asian countries such as Japan, India, China, and South Korea, it was in the range of 130–180 g CO_2/km; for countries in the Americas, such as the United States, Canada, and Mexico, it was in the range of 180–220 g CO_2/km; and for the European Union, it was 135 g CO_2/km [45]. Based on these comparisons, the CO_2 emissions intensity per car activity in Indonesia can be said to be high. Therefore, more efforts should be undertaken to significantly reduce CO_2.

3.1.6. Model Validation

Validation compares other data with the results for the provincial and national models. Looking at the standard error of results for 2010–2015, the provincial model has a standard error of estimates 0.0326, while the national model's was 0.0516. This finding demonstrates that the accuracy of the provincial model is higher than the national model. Figure 9 illustrates the comparison of energy consumption between the model results and the data from Ministry of Energy and Mineral Resources of Indonesia [1].

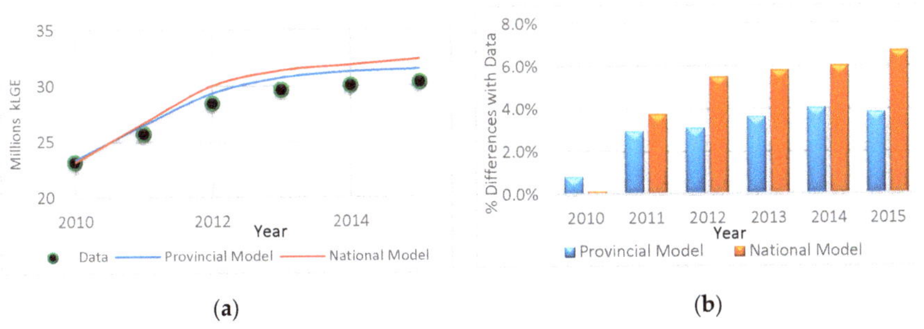

Figure 9. Comparison of energy consumption between data with model results. (**a**) Energy consumption; (**b**) Percentage.

3.2. Projection Results

3.2.1. Projection of Car Ownership

Figure 10 shows car ownership projections for provinces grouped by region. These projections show disparities among provinces. In 2015, the difference of car ownership among provinces was in the range of 3–344 vehicles/1000 people, with the average car ownership across provinces being 64 vehicles/1000 people. In 2050, the discrepancy is expected to widen, with an estimated range of 117–603 vehicles/1000 people and average car ownership across provinces at 479 vehicles/1000 people. In 2050, the smallest discrepancy is expected to appear for the Kalimantan and Sumatra regions, and the largest for the Nusa Tenggara, Maluku, and Papua regions. The provinces of Maluku and North Maluku, which are mostly situated on an archipelago, show relatively low rates of car ownership. The first province to experience car ownership saturation is DKI Jakarta, with most provinces approaching the saturated condition and a few more that are just starting to approach saturation.

Figure 11 shows a comparison of the top five provinces by number of cars. In 2015, the number of cars in Jakarta was the highest, but in 2050, Jakarta is not expected to be in the top five, because car ownership in Jakarta has already reached saturation, with the population at its maximum level. In 2050 it is also expected that approximately 50% of cars will continue to be concentrated in the Java region.

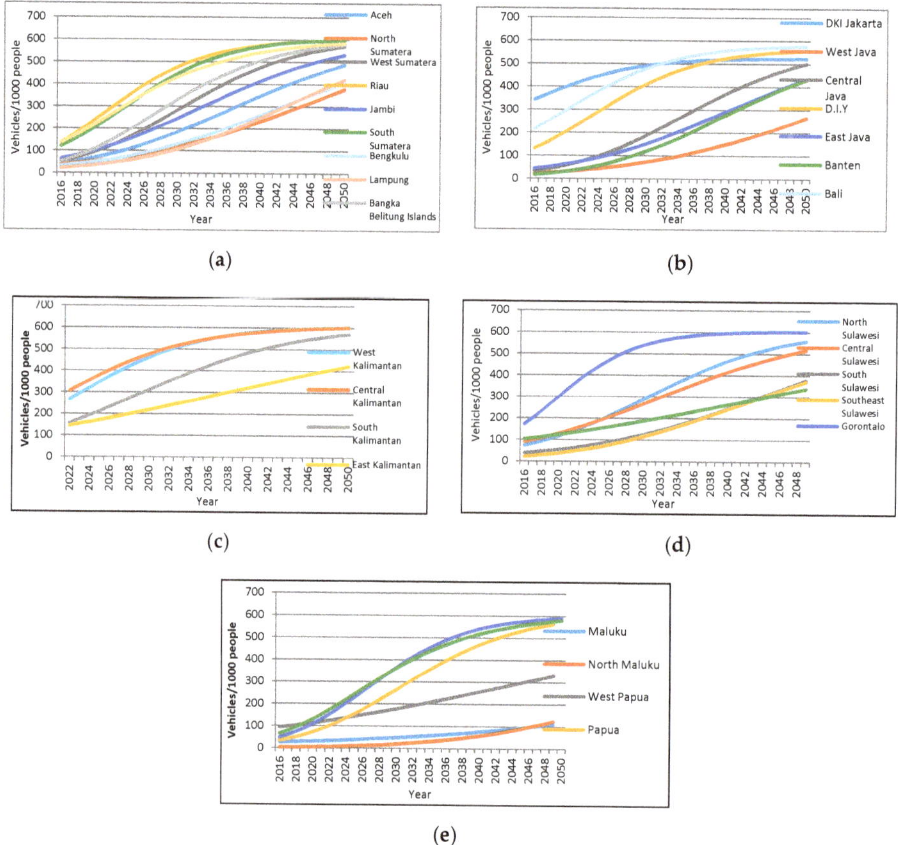

Figure 10. Projection of car ownership in (**a**) Sumatra (**b**) Java (**c**) Kalimantan (**d**) Sulawesi (**e**) Nusa Tenggara, Maluku, Papua.

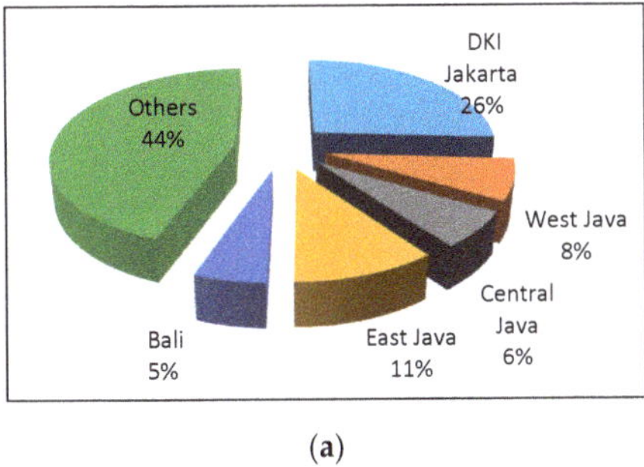

(a)

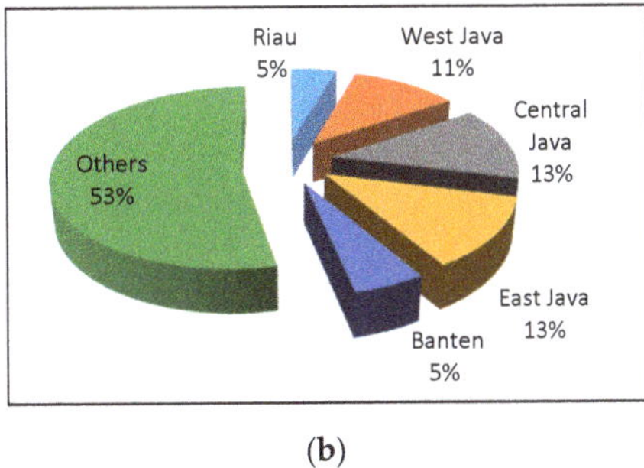

(b)

Figure 11. Comparison of the Top 5 provinces by number of cars (**a**) in 2015 (**b**) in 2050.

3.2.2. Impact of Policy Scenario

The BAU scenario is used as a reference for the other scenarios in terms of energy demand and CO_2 emissions reduction. The differences between the BAU scenario and other scenarios are in the intensity and activity of cars; therefore, fuel economy and VKT will also differ among scenarios. Fuel economy in the BAU scenario shows an improvement, as depicted in Figure 12.

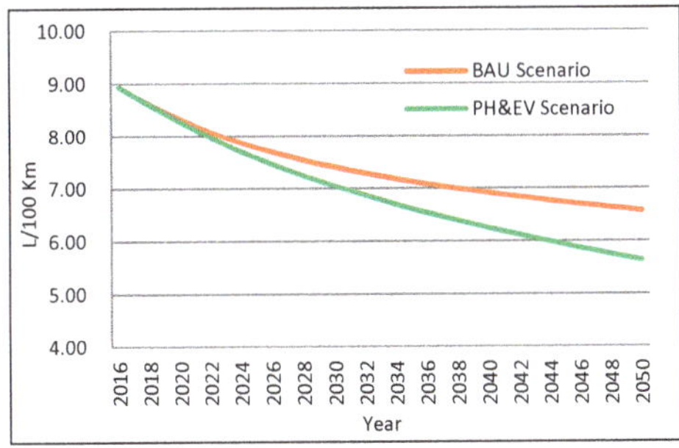

Figure 12. Projected National Fuel Economy, 2016–2050.

Fuel economy improvement in the projected BAU scenario occurs because car manufacturers are expected to improve their fuel economy regardless of the enactment of specific policies. However, this improvement in fuel economy is not as significant as in the car technology scenario. The car technology scenario leads to significant improvement in fuel economy. According to a previous study [46], fuel economy improvements can occur even if technological developments for increasing vehicle efficiency are only directed at improving fuel economy, and the performance of the vehicle remains constant. This study has analyzed possibilities in fuel economy improvement through modifications such as decreasing the weight and size of the car, in the absence of technological developments that increase the acceleration and horsepower performance [46]. These kinds of modifications are used in the assumptions of car fuel economy improvements for the car technology scenario.

The VKTs decrease slightly in the BAU scenario due to fuel price increases. Changes in fuel prices are more likely to occur as crude oil price increases, according to the crude oil price projections reported by the US Energy Information Administration [34]. Table 9 shows the *VKM* at BAU conditions for each province.

Table 9. *VKM* projection results for provinces, BAU scenario, 2016–2050 (million *VKM*).

No	Province	2016	2020	2030	2040	2050
1	Aceh	1332	1701	4128	6749	8285
2	North Sumatera	5125	6406	13,544	23,550	33,210
3	West Sumatera	2964	4406	11,910	17,758	19,007
4	Riau	6887	10,596	20,416	22,872	21,605
5	Jambi	900	1138	2534	3770	4279
6	South Sumatera	4717	6533	13,263	15,422	14,630
7	Bengkulu	499	667	1603	2829	3833
8	Lampung	2353	3192	8688	17,472	25,492
9	Bangka Belitung Islands	861	1320	3537	4797	4901
10	Riau Islands	2291	3211	6062	7035	6842
11	DKI Jakarta	12,181	12,428	15,171	14,914	13,635
12	West Java	27,079	33,387	72,478	131,456	198,790
13	Central Java	9183	14,281	42,067	73,286	87,673
14	D.I.Y	1726	2199	4119	4763	4560
15	East Java	14,251	18,439	41,458	70,599	92,029
16	Banten	7127	10,982	36,683	73,275	99,262
17	Bali	4190	4825	7407	7905	7391
18	West Nusa Tenggara	1632	3249	9989	13,504	13,306
19	East Nusa Tenggara	1875	3242	8277	10,790	10,870
20	West Kalimantan	3592	5197	10,415	11,926	11,252
21	Central Kalimantan	2523	3408	6035	6717	6347
22	South Kalimantan	2418	3428	7869	10,995	11,600
23	East Kalimantan	2561	2366	3658	4721	5423
24	North Sulawesi	1358	1772	3999	5737	6167
25	Central Sulawesi	1023	1240	2415	3366	3756
26	South Sulawesi	5192	6934	15,327	26,892	37,333
27	Southeast Sulawesi	384	574	1600	2960	4066
28	Gorontalo	1398	2089	3604	3708	3402
29	West Sulawesi	470	554	844	1095	1270
30	Maluku	357	352	525	721	950
31	North Maluku	108	163	546	1351	2604
32	West Papua	590	668	1034	1358	1591
33	Papua	1332	2258	7260	11,323	12,256
	Indonesia	130,478	173,204	388,465	615,618	777,617

The *VKM* projections in the BAU scenario show disparities among the provinces. In 2050, the five provinces with the highest *VKM* will be West Java, East Java, DKI Jakarta, Central Java, and Riau. National *VKM* is an aggregation of the *VKM* of all provinces. The comparison of national *VKM* among the different scenarios is shown in Figure 13.

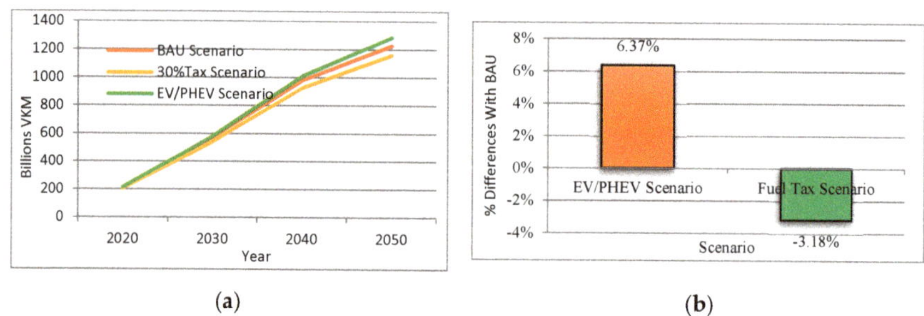

(**a**) (**b**)

Figure 13. Comparison of VKM among the scenarios: (**a**) 2016–2050, and (**b**) 2050.

Based on Figure 13a, the fuel tax scenario has the lowest value for VKM. The fuel tax scenario reduces VKM by 3.18%, while the VKM in the car technology scenario tends to be higher than in the

BAU scenario, because the significant fuel economy improvement causes the fuel cost to decrease. Consequently, this may precipitate an increase in VKM. This effect is commonly referred to as a rebound effect, such that fuel economy improvement does not reduce energy demand but instead increases it.

The energy demand projections for all provinces are shown in Table 10. The top five provinces in terms of energy demand increase are North Maluku, Southeast Sulawesi, Banten, Papua, and Lampung. These increases are caused by the growth rate of car ownership, which is influenced by a combination of α and β and also by the high VKT in preceding years. The highest energy demand is predicted to occur in 2030, because a take-off phase in levels of car ownership is expected in many provinces in that year.

Table 10. Energy demand projections for provinces, BAU scenario, 2016–2050 (LGE).

No	Province	2016	2020	2030	2040	2050
1	Aceh	119,148,323	141,523,449	303,607,934	461,464,806	540,244,217
2	North Sumatera	458,543,142	533,014,420	996,144,764	1,610,238,179	2,165,457,146
3	West Sumatera	265,192,718	366,591,863	875,942,962	1,214,204,525	1,239,355,128
4	Riau	616,226,815	881,589,123	1,501,512,434	1,563,904,525	1,408,724,650
5	Jambi	80,552,871	94,647,348	186,339,112	257,809,352	279,043,157
6	South Sumatera	422,063,720	543,568,995	975,487,176	1,054,481,349	953,949,811
7	Bengkulu	44,636,584	55,464,461	117,917,482	193,412,673	249,924,630
8	Lampung	210,516,097	265,594,157	638,973,126	1,194,670,091	1,662,176,551
9	Bangka Belitung Islands	77,063,962	109,808,609	260,164,030	328,000,084	319,551,372
10	Kepulauan Riau	204,942,989	267,180,562	445,877,246	481,031,935	446,116,423
11	DKI Jakarta	1,089,849,014	1,034,018,568	1,115,802,771	1,019,749,331	889,080,510
12	West Java	2,422,887,664	2,777,829,389	5,330,535,379	8,988,375,156	12,962,040,540
13	Central Java	821,685,057	1,188,229,370	3,093,896,625	5,010,986,595	5,716,720,151
14	D.I.Y	154,409,331	182,962,705	302,952,178	325,705,446	297,304,363
15	East Jawa	1,275,065,275	1,534,166,702	3,049,075,086	4,827,280,936	6,000,723,531
16	Banten	637,695,462	913,691,429	2,697,915,219	5,010,194,194	6,472,332,110
17	Bali	374,879,124	401,436,895	544,728,515	540,503,534	481,907,250
18	West Nusa Tenggara	146,061,501	270,335,225	734,625,142	923,360,040	867,604,144
19	East Nusa Tenggara	167,767,743	269,777,612	608,735,560	737,792,520	708,745,604
20	West Kalimantan	321,358,616	432,411,794	766,017,330	815,450,870	733,708,303
21	Central Kalimantan	225,759,254	283,570,145	443,844,722	459,285,481	413,845,972
22	South Kalimantan	216,361,886	285,199,263	578,752,860	751,783,980	756,379,095
23	East Kalimantan	229,105,517	196,859,130	269,020,170	322,828,260	353,611,885
24	North Sulawesi	121,545,236	147,408,264	294,134,362	392,257,298	402,148,817
25	Central Sulawesi	91,562,634	103,195,605	177,637,834	230,150,517	244,921,880
26	South Sulawesi	464,591,258	576,928,440	1,127,220,035	1,838,777,953	2,434,305,858
27	Southeast Sulawesi	34,396,473	47,753,362	117,644,424	202,392,077	265,101,108
28	Gorontalo	125,043,868	173,800,532	265,064,967	253,555,081	221,825,482
29	West Sulawesi	42,066,227	46,053,319	62,040,380	74,887,606	82,830,661
30	Maluku	31,965,517	29,320,690	38,589,159	49,286,590	61,960,934
31	North Maluku	9,620,306	13,548,369	40,155,478	92,387,460	169,790,491
32	West Papua	52,763,586	55,606,413	76,011,800	92,843,257	103,709,345
33	Papua	119,173,052	187,851,251	533,985,506	774,195,128	799,173,697
	Indonesia	11.674.500.823	14,410,937,457	28,570,351,767	42,093,246,828	50,704,314,820

In DKI Jakarta, the energy demand tends to be stable, even decreasing in 2050. This decrease is due to the fuel economy of cars, which continues to decline from year to year, while car ownership remains stable because of the steady population. According to the projections from the Central Bureau of Statistics of Indonesia, in 2050 DKI Jakarta's population is predicted to increase by only 14%, while the average population growth throughout all provinces will be approach 41%. This means the number of cars in DKI Jakarta cannot increase significantly. As a result, decreases in fuel economy would be able to offset the increase in VKM, while for the other provinces, the reverse situation applies. Figure 14 shows the comparison between scenarios for energy demand.

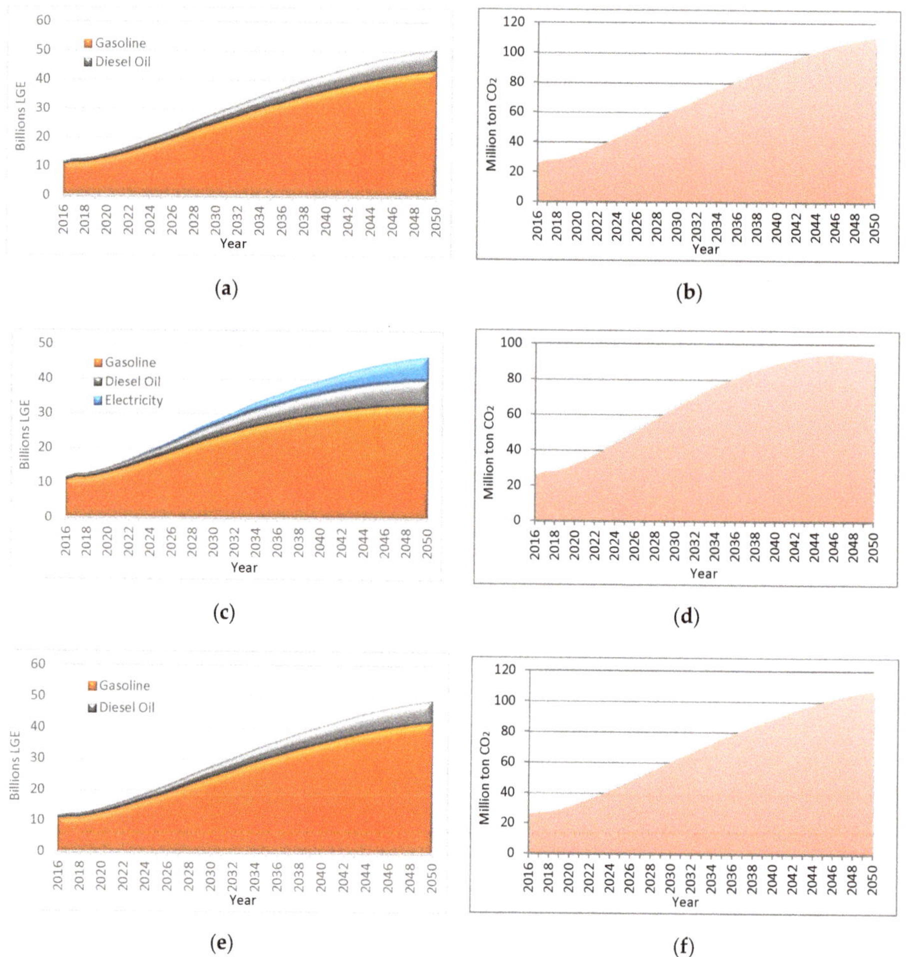

Figure 14. Results of energy demand and CO_2 emissions among scenarios. (**a**) Energy Demand (BAU Scenario); (**b**) CO_2 Emissions (BAU Scenario); (**c**) Energy Demand (Car Tech. Scenario); (**d**) CO_2 Emissions (Car Tech. Scenario); (**e**) Energy Demand (Fuel Tax Scenario); (**f**) CO_2 Emissions (Fuel Tax Scenario).

The BAU scenario projections show that in 2050, the energy demand and CO_2 emissions will reach 50 million LGE and 110 million tons, respectively. This situation is about 4.3 times higher than in 2015. Moreover, the energy demand in the car technology and fuel tax scenarios will reach 46 and 49 million LGE, while the CO_2 emissions will reach 93 and 107 million tons, respectively. Figure 15 shows the comparison of CO_2 emission reduction in 2050 among all scenarios. The highest performance in terms of CO_2 emissions reduction occurs in the car technology scenario. The car technology scenario shows greater reduction due to the sales mix of PHEV and EV reaching 50% in 2050, with the accumulated number of PHEV and EV cars reaching 17.6 million, or 18% of the total car population. Moreover, the large number of CO_2 emission reductions in the car technology scenario occurred due to significant decarbonization of the electricity generation and share technology vehichle.

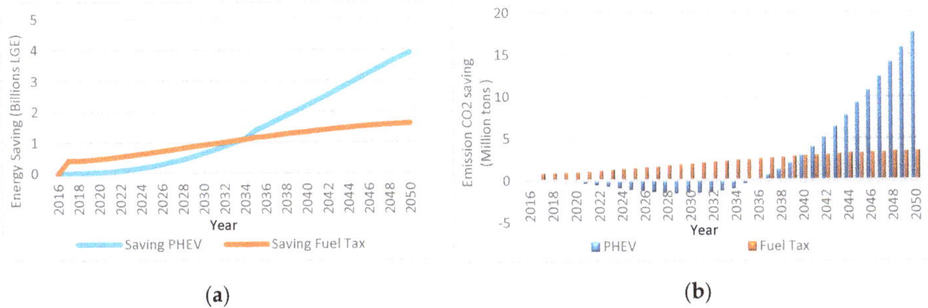

(a) (b)

Figure 15. Comparison between scenarios for energy demands and CO_2 emission savings (**a**) Energy demand savings; (**b**) CO_2 emission savings.

To realize this market penetration of PH/EV, several problems need to be overcome: limited battery car capacity, the cost of batteries, charging infrastructure, economies of scale, and the total cost of operating the PH/EV against liquid fuel car operation. The government needs to devise better strategies, including a roadmap outlining battery charging infrastructure, fiscal policies to reduce the total cost of PH/EV, in order to create a more competitive market for the PH/EV cars. Further strategy to be implemented is green incentives to increase the willingness to pay of the electric vehicle, therefore the electricity vehicle's ownership will be increased.

The fuel tax scenario reduces CO_2 emissions through VKM reduction. Since 2015, the government has eliminated subsidies, demonstrating that a fuel tax can be an effective means to control car travel. A tax of 30% could reduce CO_2 emissions by 3.18%. However, the tax regulation should take into account the people's purchasing power. Therefore, the government should increase the people's purchasing power and consider fuel price based on fuel quality. Figure 16 shows the expected CO_2 emissions disparities among provinces in 2050.

Figure 16. CO_2 emissions disparities among provinces, BAU scenario, 2050 (million ton CO_2).

The disparity of CO_2 emissions among provinces is quite striking, especially the disparity between western and eastern Indonesia. Special attention should be given to western Indonesia, then, particularly the Java region. The five provinces expected to contribute the most to CO_2 emissions by 2050 are West Java, East Java, Central Java, Banten, and South Sulawesi. The CO_2 emissions in DKI Jakarta are not expected to change much, while adjacent provinces are likely to experience high CO_2 emissions.

In 2050, the values for CO_2 emissions intensity per car activity for the BAU and car technology scenarios are 145 and 114 g CO_2/km, respectively, while the values for the fuel tax scenario are similar to those for the BAU scenario. The car technology scenario shows a significant improvement, with 15.96% lower emissions than in the BAU scenario. However, such emission reductions require a significant reduction in electricity emission factors to be near zero kg CO_2/GJ by 2050 which can be done through increasing the supply of electricity from renewable energy sources.

4. Conclusions

This study analyzes energy demand and CO_2 emissions in Indonesia in a historical situation (2010–2015) and during a projected period (2016–2050) resulting from the use of passenger cars. The results show disparities among provinces, which are mainly due to differences in GDP, population, area, and the number of cars. The historical situation shows that in 2015, the energy demand and CO_2 emissions from passenger cars amounted to 10 million LGE and 23 million tons of CO_2, respectively. In 2050, these values are expected to reach 50 million LGE and 110 million ton of CO_2, respectively, which is 4.3 times higher than that in 2015.

The five provinces with the highest CO_2 emissions in the historical situation, particularly in 2015, are West Java, East Java, DKI Jakarta, Central Java, and Banten. In 2050, the top five are predicted to be West Java, Banten, East Java, Central Java, and South Sulawesi. Therefore, special attention needs to be accorded to these provinces.

Compared to the BAU condition, the car technology and fuel tax scenarios could reduce energy demand by 7.72% and 3.18% and CO_2 emissions by 15.96% and 3.18%, respectively. The car technology scenario requires certain policies in order to achieve the reduction in CO_2 emissions, such as car economy labeling and fuel economy standards. Economy labeling is an obligation for car manufacturers and dealers to provide information on car fuel economy, while fuel economy standards are enacted by limiting car fuel economy based on the vehicle's class and intended purposes. In addition, this scenario requires a significant reduction in electricity emission factors to be 27.8 kg CO_2/GJ by 2050.

The projected fuel tax scenario could reduce CO_2 emissions by 3.18% in 2050. This scenario could be realized by imposing higher taxes in order to limit car activity. The higher the tax, the lower the CO_2 emissions; however, the imposition of fuel tax should also consider the ability of people to buy fuel, which is in line with GDP per capita.

The model for energy demand and CO_2 emissions of passenger cars at the provincial level can improve the accuracy of the analysis when aggregated to the country level, which is proven by model validation.

The current study's results could be used by provincial governments as an overview of energy and CO_2 emissions contributions by passenger cars. Furthermore, some scenarios have been given to illustrate possibilities for CO_2 emissions reduction. Special attention should be given to provinces which are the largest contributors to the current problem and also to those expected to experience significant increases in CO_2 emissions in the future.

Author Contributions: Conceptualization, W.W.P.; Methodology, Q.F.E.; Validation, W.W.P., and M.S.; Formal Analysis, Q.F.E.; Investigation, Q.F.E.; Data Curation, Q.F.E.; Writing-Original Draft Preparation, Q.F.E.; Writing-Review & Editing, M.S., N.R.; Visualization, Q.F.E.; Supervision, W.W.P.; Project Administration, W.W.P.; Funding Acquisition, W.W.P.

Funding: This research was funded by the Directorate for Research and Public Services (DRPM) Universitas Indonesia, grant number Hibah Publikasi Artikel di Jurnal Internasional Kuartil Q1 dan Q2 (Q1Q2) NKB-0328/UN2.R3.1/HKP.05.00/2019.

Conflicts of Interest: The authors declare no conflict of interest.

Nomenclature

BOE	Barrel oil equivalent
ASIF	Activity Structure Intensity Fuel
LCGC	Low Cost Green Car
CO	car ownership
*CO**	saturated car ownership
GDPP	GDP per capita
D	population density
FE	fuel economy
FC	fuel cost
VKT	vehicle kilometer traveled
VKM	vehicle kilometers
ϵ	elasticity
EF	emission factor
E	energy demand
G	CO_2 emission

References

1. Ministry of Energy and Mineral Resources Republic of Indonesia. *Handbook of Energy & Economic Statistics of Indonesia, Head of Center for Energy and Mineral Resources Data and Information*; Ministry of Energy and Mineral Resources Republic of Indonesia: Jakarta, Indonesia, 2016.
2. Erahman, Q.F.; Purwanto, W.W.; Sudibandriyo, M.; Hidayatno, A. An assessment of Indonesia's energy security index and comparison with seventy countries. *Energy* **2016**, *111*, 364–376. [CrossRef]
3. Dargay, J.; Gately, D. Vehicle ownership to 2015: Implications for energy use and emissions. *Energy Policy* **1997**, *25*, 1121–1127. [CrossRef]
4. Dargay, J.; Gately, D. Income's effect on car and vehicle ownership, worldwide: 1960–2015. *Transp. Res. Part A Policy Pract.* **1999**, *33*, 101–138. [CrossRef]
5. Dargay, J.; Gately, D. Modelling global vehicle ownership. In Proceedings of the Ninth World Conference on Transport Research, Seoul, Korea, 22–27 July 2001.
6. Dargay, J.; Gately, D.; Sommer, M. Vehicle ownership and income growth, worldwide: 1960–2030. *Energy J.* **2007**, *28*, 143. [CrossRef]
7. Leaver, L.; Samuelson, R.; Leaver, J. Potential for low transport energy use in developing Asian cities through compact urban development. In Proceedings of the Third IAEE Asian Conference, Kyoto, Japan, 20–22 February 2012.
8. Wu, T.; Zhao, H.; Ou, X. Vehicle ownership analysis based on GDP per capita in China: 1963–2050. *Sustainability* **2014**, *6*, 4877–4899. [CrossRef]
9. Ministry of Environment Republic of Indonesia. President Regulation No.61/2011. 2011. Available online: http://www.bappenas.go.id/files/6413/5228/2167/perpres-indonesia-ok__20111116110726__5.pdf (accessed on 10 December 2017).
10. Ministry of Environment Republic of Indonesia. President Regulation No.71/2011. 2011. Available online: http://peraturan.go.id/perpres/nomor-71-tahun-2011-11e44c4f4f4b3a90bde2313232303234.html (accessed on 10 December 2017).
11. International Energy Agency. *Energy Technology Perspectives*; International Energy Agency: Paris, France, 2012.
12. International Energy Agency. *Technology roadmap: Fuel Economy of Road Vehicles*; International Energy Agency: Paris, France, 2012.
13. Zhang, M.; Mu, H.; Li, G.; Ning, Y. Forecasting the transport energy demand based on PLSR method in China. *Energy* **2009**, *34*, 1396–1400. [CrossRef]
14. Lu, I.J.; Lewis, C.; Lin, S.J. The forecast of motor vehicle, energy demand and CO_2 emission from Taiwan's road transportation sector. *Energy Policy* **2009**, *37*, 2952–2961. [CrossRef]
15. Chai, J.; Lu, Q.Y.; Wang, S.Y.; Lai, K.K. Analysis of road transportation energy consumption demand in China. *Transp. Res. Part D Transp. Environ.* **2016**, *48*, 112–124. [CrossRef]

16. Bhattacharyya, S.C. *Energy Economics: Concepts, Issues, Markets and Governance*; Springer Science & Business Media: Berlin, Germany, 2011.
17. Eom, J.; Schipper, L. Trends in passenger transport energy use in South Korea. *Energy Policy* **2010**, *38*, 3598–3607. [CrossRef]
18. Ma, L.; Liang, J.; Gao, D.; Sun, J.; Li, Z. The Future Demand of Transportation in China: 2030 Scenario based on a Hybrid Model. *Procedia Soc. Behav. Sci.* **2012**, *54*, 428–437. [CrossRef]
19. Baptista, P.C.; Silva, C.M.; Farias, T.L.; Heywood, J.B. Energy and environmental impacts of alternative pathways for the Portuguese road transportation sector. *Energy Policy* **2012**, *51*, 802–815. [CrossRef]
20. Ko, A.; Myung, C.L.; Park, S.; Kwon, S. Scenario-based CO2 emissions reduction potential and energy use in Republic of Korea's passenger vehicle fleet. *Transp. Res. Part A Policy Pract.* **2014**, *59*, 346–356. [CrossRef]
21. Widyaparaga, A.; Sopha, B.M.; Budiman, A.; Muthohar, I.; Setiawan, I.C.; Lindasista, A.; Soemardjito, J.; Oka, K. Scenarios analysis of energy mix for road transportation sector in Indonesia. *Renew. Sustain. Energy Rev.* **2017**, *70*, 13–23.
22. Central Statistics Bureau Indonesia. Statistik Nasional. 2013. Available online: http://www.bps.go.id/menutab.php?tabel=1&kat=2&id_subyek=17 (accessed on 25 November 2017).
23. Gabungan Industri Kendaraan Bermotor. Gaikindo Website. 2017. Available online: www.gaikindo.or.id (accessed on 25 November 2017).
24. Central Statistics Bureau Indonesia. Proyeksi Penduduk Menurut Provinsi. 2017. Available online: https://www.bps.go.id/linkTabelStatis/view/id/1274 (accessed on 25 November 2017).
25. Central Statistics Bureau Indonesia. *Statistik Indonesia: Statistical Yearbook of Indonesia*; Biro Pusat Statistik: Jakarta, Indonesia, 2016.
26. Ministry of Industry Republic of Indonesia. *Minister of Industry Regulation No.33/M-IND/PER/7/2013*; Ministry of Industry Republic of Indonesia: Jakarta, Indonesia, 2013.
27. Ministry of Industry Republic of Indonesia. Press Conference. 2013. Available online: http://www.kemenperin.go.id/artikel/6775/Menperin-Keluarkan-Peraturan-Mobil-LCGC (accessed on 10 December 2017).
28. Automobile Catalog. 2012 Daihatsu Xenia Xi (Model for Asia) Specifications & Performance Data Review. 2017. Available online: http://www.automobile-catalog.com/car/2012/1223810/daihatsu_xenia_xi.html (accessed on 17 October 2017).
29. Auto-data.net. Toyota Innova Fuel Economy. 2017. Available online: https://www.auto-data.net/en/?f=showCar&car_id=3435 (accessed on 17 October 2017).
30. Australia, C.O. GreenVehicleGuide. Available online: https://www.greenvehicleguide.gov.au/ (accessed on 20 October 2017).
31. International Energy Agency. *International Comparison of Light-Duty Vehicle Fuel Economy: Evolution Over 8 Years from 2005 to 2013*; International Energy Agency: Paris, France, 2014.
32. International Energy Agency. *Transport, Energy and CO2: Moving Towards Sustainability*; International Energy Agency: Paris, France, 2009; p. 44.
33. International Energy Agency. *Fuel Economy Policies Implementation Tool (FEPIT)*; International Energy Agency: Paris, France, 2015.
34. U.S. Energy Information Administration. Analysis & Projections of Petroleum Prices. 2017. Available online: https://www.eia.gov/analysis/projection-data.cfm#annualproj (accessed on 25 October 2017).
35. Stratas Advisors. Global Refining Outlook: 2016–2035. 2016. Available online: https://stratasadvisors.com/~/media/Marketing%20Images/Newsroom%20Items/GlobalRefiningOutlook_Webinar_March10_2016.pdf?la=en (accessed on 15 October 2017).
36. Ministry of Energy and Mineral Resources Republic of Indonesia. *MEMR Regulation No.1713 K/12 MEM/2012*; Ministry of Energy and Mineral Resources: Central Jakarta, Indonesia, 2012.
37. Ministry of Environment Republic of Indonesia. Guidelines of National Greenhouse Gas Inventories. 2012. Available online: http://www.kemenperin.go.id/download/11219 (accessed on 10 December 2017).
38. Ministry of Energy and Mineral Resources Republic of Indonesia. Faktor Emisi Pembangkit Listrik Sistem Interkoneksi. 2016. Available online: http://www.djk.esdm.go.id/pdf/Faktor%20Emisi%20Pembangkit%20Listrik/Faktor%20Emisi%20Pembangkit%20Listrik%20Tahun%202011-2014.pdf (accessed on 10 December 2017).
39. International Energy Agency. *Transport energy efficiency*; International Energy Agency: Paris, France, 2010.

40. Secretariat of Cabinet Republic Indonesia. *President Regulation No.22/2017*; Secretariat of Cabinet Republic Indonesia: Jakarta, Indonesia, 2017.
41. International Energy Agency. *Technology Roadmap: Electric and Plug-in Hybrid Electric Vehicles*; International Energy Agency: Paris, France, 2011.
42. Preece, R. Excise taxation of key commodities across South East Asia: A comparative analysis ahead of the ASEAN Economic Community in 2015. *World Cust. J.* **2012**, *6*, 3–16.
43. Ministry of Energy and Mineral Resources Republic of Indonesia. MEMR Regulation No.32/2008. 2008. Available online: http://prokum.esdm.go.id/permen/2008/permen-esdm-32-2008.pdf (accessed on 10 December 2017).
44. Ministry of Energy and Mineral Resources Republic of Indonesia. MEMR Regulation No.25/2013. 2013. Available online: http://prokum.esdm.go.id/permen/2013/Permen%20ESDM%2025%202013.pdf (accessed on 10 December 2017).
45. Façanha, C.; Blumberg, K.; Miller, J. *Global Transportation Energy and Climate Roadmap*; The International Council for Clean Transportation: Washington, DC, USA, 2012.
46. Cheah, L.; Heywood, J. Meeting, U.S. passenger vehicle fuel economy standards in 2016 and beyond. *Energy Policy* **2011**, *39*, 454–466. [CrossRef]

MDPI
St. Alban-Anlage 66
4052 Basel
Switzerland
Tel. +41 61 683 77 34
Fax +41 61 302 89 18
www.mdpi.com

Energies Editorial Office
E-mail: energies@mdpi.com
www.mdpi.com/journal/energies

www.ingramcontent.com/pod-product-compliance
Lightning Source LLC
LaVergne TN
LVHW070845170726
843515LV00005B/579

* 9 7 8 3 0 3 6 5 0 0 1 5 7 *